105 Advances in Polymer Science

Recent Trends
in Radiation
Polymer Chemistry

Editor: S. Okamura

With contributions by
T. Ichikawa, I. Kaetsu, M. Ogasawara,
S. Tagawa, H. Yamaoka, H. Yoshida

With 74 Figures and 12 Tables

Springer-Verlag
Berlin Heidelberg GmbH

ISBN 978-3-662-14945-4   ISBN 978-3-540-47292-6 (eBook)
DOI 10.1007/978-3-540-47292-6

Springer-Verlag Berlin Heidelberg 1993
Originally published by Springer-Verlag Berlin Heidelberg New York in 1993
Softcover reprint of the hardcover 1st edition 1993

Library of Congress Catalog Card Number 61-642

Typesetting: Macmillan India Ltd., Bangalore-25

02/3020 5 4 3 2 1 0 Printed on acid-free paper

Editors

Table of Contents

Introductory Remarks

Seizo Okamura

General planning of this volume came from a discussion by the authors on the continuity of their researches with respect to the past, present and the future investigations of all the scientists involved.

The discipline of radiation polymer chemistry has long been influenced by three different fields; radiation physics, radiation chemistry and polymer science in theory and in their applications.

Early developments of this field in our country [1, 2] began almost thirty years ago with research in which polymer science was earnestly and progressively investigated from the view of both physics and chemistry. Tremendous numbers of results have accumulated over these three decades.

However, for approximately two decades, some tendencies have appeared among the public attitude toward industrial applications of ionizing radiation. They feel that there is a risk to humans and this feeling is especially prevalent in our country, because of Hiroshima and Nagasaki. Thus industrial approaches using ionizing radiation for obtaining new products and processes stagnated and slowed down, but conversely basic research was activated.

In this volume, we present five articles – two papers on the behavior of free radicals and active intermediates produced by irradiation in the polymeric matrix are studied; one article is on radiation-induced polymerization as a process for biomedical application, and finally there are two papers concerned with the radiation effects on polymeric materials from ion beams and in relation to their use for fusion reactors.

Some new trends can be recognized in the points such as the interaction of short-lived active species in some spatial distributions measured by spin echo and pulse radiolysis methods. The application of polymers for drug-delivery systems is here discussed with reference to low temperature radiation polymerization techniques. Ion beam irradiation of polymers is also reviewed for which further research is becoming important and attractive for so-called LET effects and high density excitation problems. In the applied fields the durable polymers used in strong and dense irradiation environments at extremely low temperature are here surveyed in connection with their use in nuclear fusion facilities.

Even in this rather limited volume of the Series, it is possible to learn some conceptional points of view in the new and future trends of radiation polymer chemistry both in its basic and applied forms.

References

1. Okamura S (1989) A short history of applied radiation polymer chemistry in Japan. In: Kroh J (ed) Early developments in radiation chemistry, Royal Soc. Chem., London, 321
2. Okamura S (ed) (1990) Thirty-year history of radiation polymer chemistry in Japan (in Japanese), Commemoration Committee, Osaka

Electron Spin Echo Studies of Free Radicals in Irradiated Polymers

H. Yoshida and T. Ichikawa
Faculty of Engineering, Hokkaido University, Kita-ku, Sapporo 060, Japan

Studies of the utilization of the electron spin echo method for elucidating the nature and behavior of free radicals in γ- irradiated polyethylene and poly(methyl methacrylate) are reviewed to demonstrate the importance and usefulness of this method generally for studying the effects of radiation on polymers. The paramagnetic relaxation was found to be largely due to the dipolar interaction between radicals and generally follow non-exponential kinetics. The dependence of the relaxations on the molecular motion and, therefore, on the site of the radicals is utilized for the relaxation-resolved electron spin resonance measurement. This new technique was able to discriminate the overlapped ESR spectra due to coexisting free radicals from each other and facilitated the elucidation of the behavior of the radicals. The dependence of the relaxations on the radiation dose or on the thermal annealing of radicals gave information on the local spatial distribution of the radicals, or the structure of a radiation induced spur. The radicals were found to be generated and trapped pairwisely in a spur. The average separation distance between the radicals in a pair was estimated to be 2.6 and 3.1 nm for polyethylene irradiated at 77 K and poly(methyl methacrylate) irradiated at room temperature. These electron spin echo results were discussed in terms of radiation-induced reactions leading to the crosslinking or the degradation of the polymers.

1 Introduction

Very primary events in the chemical effect of radiations on matter are excitation and ionization of molecules, which result in the formation of neutral free radicals and radical ions. These reactive species play vital roles in the radiation-induced chemical reactions. As they are paramagnetic with an unpaired electron, electron spin resonance (ESR) spectroscopy has been a useful method for elucidating the mechanism of radiation-induced reactions in solid matter where radical species can be trapped temporarily. Since the early days of the chemical application of ESR, this method has been applied very often to the identification and quantification of free radicals in polymers irradiated by radiation [1]. This is probably because, from the view-point of fundamental research, a variety of free radicals are readily trapped in solid polymers and, from the view-point of applied research, these free radicals have close correlation with radiation-induced crosslinking and degradation of polymers.

In the conventional ESR method using continuous microwave radiation (cw ESR), the identification and quantification of radical species are made from the spectral shape and the spectral intensity, respectively, under the condition of a low enough level of microwave power incident to the sample cavity. If the power level is too high, the structure of ESR spectra becomes broadened and obscure and the intensity of the spectra is no longer proportional to the radical concentration (power saturation effect). Care is usually taken to avoid these effects in cw ESR measurements.

The broadening and saturation at a high power level are closely related to paramagnetic relaxation of radical species. The information on the paramagnetic relaxation has not been widely utilized in cw ESR spectroscopy of irradiated polymers because of experimental difficulties and indirect implication of results. However, there have been some studies on the paramagnetic relaxation of the radicals in polymers by means of the cw-ESR method. In 1964, Bullock and Sutcliffe [2], and Yoshida et al. [3] found that the local concentration of free radicals in γ-irradiated polyethylene and poly(methyl methacrylate) estimated from the power saturation behavior was higher than the bulk average concentration by more than ten times. This might be the first experimental proof of the heterogeneous spatial distribution of radiation effects in condensed matters. The heterogeneous distribution of free radicals in polyethylene were later studied by the ESR saturation method by other groups [4, 5].

The ionization of a molecule and the rupture of a chemical bond by ionizing radiation necessarily result in the pairwise formation of radical species. The pairwise correlation of radical species will be more or less retained in solid polymers where the radical migration is restricted. This heterogeneity of spatial distribution of radical species affects the radiation chemistry of polymers. Another source of spatial heterogeneity is the heterogeneous deposition of radiation energy [6, 7]. Low LET radiations such as γ-rays produce an ensemble of isolated spurs. Each spur is composed of a few ion-pairs and/or radical pairs. High LET radiations such as α-rays produce overlapped spurs called tracks. This kind of heterogeneity also affects radiation chemistry, because the reaction of radical species is highly dependent on their local concentration. Therefore, the examination of the spatial heterogeneity is important for elucidating radiation-chemical reactions. The spatial heterogeneity affects the paramagnetic relaxation rate through electron spin-spin interactions, so that the analysis of the paramagnetic relaxation gives a means of examining the spatial heterogeneity of radical species.

The electron spin echo (ESE) method is a kind of time-domain ESR using pulsed microwave radiation and it directly observes the relaxation behavior of electron spins. Since the construction of an ESE spectrometer was first reported by Kaplan early in 1962 [8], the ESE method has become more and more popular slowly but steadily. Owing to the progress in microwave technology and to the continual effort of pioneering workers such as Tsvetkov, Mims, Kevan and others [9, 10]. the ESE method has been extensively developed and improved. It is now considered to be requisite for studying the paramagnetic relaxation of radical species.

The ESE method opens new horizons for ESR spectroscopy. It is used not only for measuring the relaxation kinetics but also for discriminating overlapped ESR spectra from each other based on the difference in relaxation rate. The discrimination is also attained in principle by using the cw ESR saturation method, but it is actually difficult because of the line broadening effect. The ESE method is also used for detecting very weak hyperfine and superhyperfine interactions which cannot be detected by the cw ESR method. The ESE envelope signal shows a periodic change in intensity called the nuclear modulation effect, when an electron spin is surrounded by weakly-interacting nuclear spins. Analysis of this effect determines a local geometrical structure [9] such as the solvation structure of a paramagnetic entity. The ESE method can be used for detecting short-lived transients because the ESE detection system is in principle composed of components with a fast time-response [11].

The present paper aims at giving a review of the studies utilizing the ESE method for the investigation of the free radicals in polyethylene and poly(methyl methacrylate) irradiated by γ-rays. These two polymers are known to be typical examples of radiation-crosslinking and radiation-degradable polymers. The main concern in the study of polyethylene is the pairwise formation of alkyl radicals in relation to the radiation-induced crosslinking. In the study of poly(methyl methacrylate), it is the discrimination of the overlapped ESR spectra for the elucidation of the radiation-induced degradation reactions. We hope these case studies may clearly demonstrate the importance and usefulness of the ESE method generally for investigating the radiation effects on polymers.

2 Paramagnetic Relaxation of Free Radicals

2.1 Paramagnetic Relaxation and Electron Spin Echo

Let us begin by outlining the theory on paramagnetic relaxation. Although it has repeatedly been a subject of previous excellent treatises [9, 12], we dare to repeat it again here because it is an essential basis for using the ESE method to study the nature and behavior of free radicals.

In a static magnetic field, each unpaired electron-spin of a free radical precesses about the axis parallel or antiparallel to the magnetic field (z-direction). The quantum state of the electron spin is expressed by α or β corresponding to these precessions. The population of the β spin state is larger than that of the α state by a Boltzman factor under the thermal equilibrium condition, so that the spin system has a total magnetization along the z-axis. The x- and y-components of the magnetic moments of electron spins are cancelled out under the thermal equilibrium condition because of the incoherence of the precession motion.

Any magnetic interaction disturbs the thermal equilibrium spin state. On the application of a microwave pulse of the resonance frequency to the spin system, an oscillating magnetic field is effectively exerted on the xy-plane, whereby the initial z-magnetization is forced to rotate toward the xy-plane. As soon as the microwave pulse is turned off, the spin system begins to precess coherently. If there are some magnetic interactions perturbing the spin system, they tend to restore the spin magnetization to the initial thermal equilibrium.

The restoration process of the spin magnetization toward the thermal equilibrium is called paramagnetic relaxation. It can be divided into two categories: longitudinal relaxation and transverse relaxation. The z-component of the total magnetization is restored by the former relaxation, while the coherence in precessing on the xy-plane is destroyed by the latter relaxation.

For visual understanding of these relaxations, it is convenient to demonstrate the motion of the spin magnetization by using a rotating frame (rotating about the z-axis) instead of the laboratory frame, as shown in Fig. 1. The oscillating magnetic field can be regarded as the superposition of two magnetic fields rotating counterwise to the microwave frequency. The rotating frame is fixed on one of the rotating fields. An observer on the rotating frame then sees the microwave magnetic field as if it acts as a static magnetic field on, say, the x-axis of the rotating frame. The external magnetic field on the z-axis is cancelled out with the rotation vector under the resonance condition (the precession frequency is equal to the microwave frequency ω_0). After the application of a microwave pulse of a proper duration, the total magnetization rotates by 90° and becomes directed on the y-axis. This corresponds to the increase in the population in the α spin state. The spin systems therefore tends to transfer the excess energy to the surroundings and be restored to thermal equilibrium. This restoration of the z-magnetization is the longitudinal relaxation.

90° PULSE   180° PULSE   ECHO

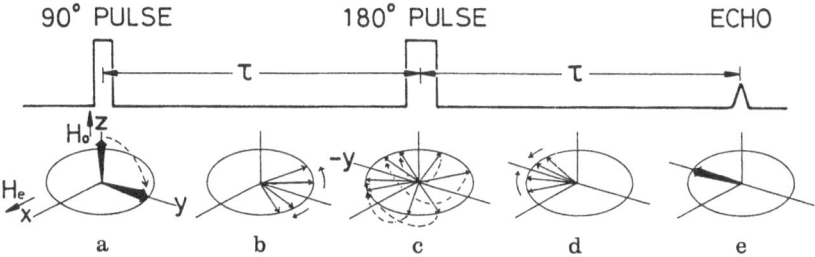

Fig. 1a–e. The vector model for paramagnetic relaxation in the rotating coordinate system showing the principle of the two-pulse spin echo. The static field H_0 and the rotating field are parallel to the z-axis and the x-axis, respectively. (**a**) The electron spins initially precess incoherently about the z-axis. The magnetic field of the first pulse, H_e, rotates the initial z-magnetization to the y-axis. (**b**) The spins dephase as soon as the first pulse is removed, so that the net magnetization on xy- plane decreases (free induction decay. (**c**) The second pulse (delayed from the first one by τ) rotates the spins about the x-axis by 180° and hence reverse the phases. (**d**) The spins begin to rephase together. (**e**) The maximum rephasing occurs and the spin echo is observed at 2τ

Even in the absence of the longitudinal relaxation, the total magnetization deoes not stay on the y-axis after the 90° pulse. The magnetization of each electron spin becomes more and more spread out on the xy-plane as the time elapses after the 90° pulse. Because the precession frequency of the excited on-resonant spins (called A spins) is not exactly the same and has some distribution depending on the intensity and the width of the microwave pulse and the shape of the ESR spectrum. The total magnetization, or the vector sum of each magnetization, quickly vanishes after the microwave pulse. This kind of motion of the total magnetization is called free induction decay.

Application of a second microwave pulse with a proper duration (180° pulse) along the x-direction at time τ after the 90° pulse rotates each magnetization about the x-axis by 180°. Suppose the precession frequency of an A spin is ω, the phase angle of the A spin with respect to the y-axis is $(\omega - \omega_0)\tau$ and $-(\omega - \omega_0)\tau - \pi$ just before and after the application of the 180° pulse, respectively. The phase angle at τ after the 180° pulse is then given by $-(\omega - \omega_0)\tau - \pi + (\omega - \omega_0)\tau = \pi$, which is independent of ω. The total magnetization is therefore refocussed on the $(-y)$-axis and the burst of the microwave energy is observed at τ after the 180° pulse. This is called an electron spin echo (ESE).

The ESE intensity does not depend on τ if the precession frequency of each electron spin does not fluctuate during the time interval 2τ. However it does more or less fluctuate under the influence of surrounding spins etc., the precession of the spins looses coherence. The refocussing becomes more and more imperfect and the ESE intensity becomes weaker and weaker as the time elapses after the 90° pulse. This behavior of the spin magnetization is called phase relaxation. Both the free induction decay and the phase relaxation are called transverse relaxation.

An electron spin interacts not only with the external magnetic field but also with surrounding electron and nuclear spins. Because of the coupling with the nuclear spins, the spin state of a radical species can no longer be defined only by the electron spin state. Instead, it is defined by the quantum state of the electron spin and that of the nuclear spins as well. The ESR spectrum of the radical species is broadened due to the difference of these interactions: this is called inhomogeneous broadening. The spectrum is further broadened by the finite lifetime of each spin quantum state: this is called homogeneous broadening. The former and the latter broadening effects cause the free induction decay and the phase relaxation, respectively.

In the above explanation, we assumed ideal 90° and 180° pulses as the excitation and refocussing pulses. The ideal pulses are, however, not necessary for observing the ESE, because any magnetization vector can be decomposed into z- and xy-components. As a matter of fact, the ideal pulses can never be obtained experimentally. The rotation angle depends on the precession frequency of A spins and the intensity of the rotating magnetic field. Since the precession frequency is not exactly the same for all the A spins and the field intensity depends on the location of A spins in the sample, the rotation angle differs radical by radical. Nevertheless, we use the terms 90° and 180° pulses in the following section because these terms are convenient for visualizing the motion of magnetization during the application of the microwave pulses.

2.2 Longitudinal Relaxation

Longitudinal relaxation is further divided into two categories: relaxation with and without change in the quantum state of the electron spins under observation (A spins). The longitudinal relaxation accompanying the change in the quantum state of the A spins is induced by either of the following two interactions.

1) Electron spin-lattice interaction. The transition between α and β spin states takes place by the interaction between the A spins and the lattice vibration of surrounding molecules. The excess energy of the A spin system is transferred to the surrounding molecules to induce lattice vibration. This relaxation takes place even for an isolated electron spin having no interaction with other electron spins at all.

2) Electron spin-electron spin interaction. The transition betwen α and β spin states takes place by the interaction between the A spins and the surrounding off-resonant spins (called B spins). The most important process in this type of the relaxation is cross relaxation. In the cross relaxation, the excess energy of the A spin system is resonantly transferred to the surrounding B spins through a flip-flop process. The relaxation rate depends on either the distance betwen the A and B spins or the number of the B spins surrounding an A spin. It is this relaxation mechanism which provides us with a means for studying the local spatial distribution of radical species.

The longitudinal relaxation without change of the electron spin state is called spectral diffusion. This is caused by the mixing of A and B spins due to the fluctuation of the resonance frequency of each spin. When the electron spin is coupled with nuclear spins, the molecular motion results in the fluctuation of the hyperfine field. The A spins then migrate into the spectral region of the B spins, the vice versa. This kind of mixing causes the longitudinal relaxation at the spectral region of the A spins, because the B spins have not become excited and are cold. The spectral diffusion is also induced by g anisotropy, quadrupole interaction, etc. The spectral diffusion provides us with a means for elucidating the molecular motion of radical species. The spectral diffusion indirectly affects the cross relaxation because this causes the overlapping of the ESR spectra of A and B spins. The cross relaxation is a resonant process and needs the spectral overlapping.

The longitudinal relaxation is often called spin-lattice relaxation and the rate of the longitudinal relaxation is often expressed by the inverse of the spin-lattice relaxation time T_1. However, this seems to be somewhat misleading. T_1 itself presumes the relaxation following the exponential law, though the exponential relaxation is seldom observed for radical species in solids. The strict meaning of the spin-lattice relaxation is the energy transfer from the A spins to phonons or lattice vibrations through the fluctuation of the spin Hamiltonian due to the lattice vibrations. Generally the longitudinal relaxation is the superposition of several relaxation processes including "pure" spin-lattice relaxation. It is usually a difficult task to extract the spin-lattice relaxation from the observed longitudinal relaxation.

The above-mentioned types of relaxation and the mechanisms involved are summarized in Table 1. Dependence of the relaxation rate on the radical concentration is also given.

Table 1. Mechanisms of paramagnetic relaxations of radical species and their dependence on radical concentration

Type of Relaxation and Interaction	Mechanism	Concentration Dependence
Longitudinal Relaxation		
A spin-Lattice	Energy transfer from A spin to lattice	No
A spin-B spin (Cross Relaxation)	Energy transfer from A spin to B spins	Yes
Spectral Diffusion	Mixing of A spins and B spins	No
Transverse Relaxation		
1) Phase Relaxation		
A spin-B spin	Fluctuation of B spin quantum state	Yes
A spin-A spin (Instantaneous Diffusion)	Change in quantum state of A spin by microwave	Yes
A spin-Nuclear spin	Fluctuation of A spin quantum state by molecular motion and nuclear spin flip-flop	No
2) Free Induction Decay	Static distribution of precession frequency	No

2.3 Transverse Relaxation

Transverse relaxation is caused by the distribution and fluctuation of the resonance frequency of the A spins. The distribution-induced relaxation is called free induction decay. The free induction decay curve is the Fourier transform of the spectral shape of the A spins. This spectral shape depends on the intensity and the pulse width of the incident microwave, when the total width of ESR spectrum is large as is the case for radical species in solids. Therefore, the analysis of the free induction decay curve gives no information on the nature of radical species in solids unless the pulse width is narrow enough to cover the entire ESR spectrum.

The fluctuation of the precession frequency for the A spins causes phase relaxation (phase memory decay). The diffusion of the ESR spectrum out of the spectral region of the A spins causes the longitudinal relaxation, whereas the diffusion within the spectral region of the A spins, together with that out of the spectral region, causes the phase relaxation. The spectral diffusion or the change of the precession frequency is induced by the fluctuation of the magnetic interactions with electron spins and/or nuclear spins surrounding the A spins. The analysis of the phase relaxation therefore gives information on the dynamics of the magnetic interaction between radical species and the surroundings.

Although it is very difficult to distinguish between the free induction decay and the phase relaxation by the conventional cw ESR method, it is quite easy to observe them separately by the ESE method [9, 10, 12], as described in detail in the following section. We will concern ourselves mainly with the phase relaxation hereafter. The phase relaxation is mainly caused by the following interactions:

1) A spin-B spin interaction. The precession frequency of an A spin varies because of the fluctuation of the quantum states of surrounding B spins. The rate of the phase relaxation increases with the concentration of B spins.

2) A spin-A spin interaction. The fluctuation of the precession frequency is also induced by the microwave pulses used for the excitation and the refocussing: the microwave pulses induce the transition betwen α and β spin states, so that the magnetic interaction of a particular A spin with other A spins is changed instantaneously during the second microwave pulse of the ESE measurements. The relaxation process due to the thus-created fluctuation is called instantaneous diffusion. The relaxation rate due to the instantaneous diffusion depends on the distance between the A spins and the number of the A spins. Because the concentration of the A spin depends on the intensity of the microwave pulse, the rate of the instantaneous diffusion also depends on the intensity of the microwave pulse: this process can be eliminated in the ESE experiments by lowering the power of the microwave pulses. When the instantaneous diffusion is the dominant process in the phase relaxation, this provides us with a means of studying the local spatial distribution of radical species [13].

3) Electron spin-nuclear spin interaction. The precession frequency of an A spin is changed by the fluctuation of its hyperfine field. If we regard the nuclear spins as B spins, the relaxation process is similar to the case for the electron spin-electron spin interaction mentioned above. The change of the hyperfine field is induced by the molecular motions of the radical species and surrounding molecules. The change is also induced by the change of spin states of weakly-interacting nuclei due to the nuclear spin flip-flop. The nuclear spin flip-flop scarcely affects the longitudinal relaxation, since the width of the fluctuation is generally much narrower than the spectral width of the A spins.

The transverse relaxation is often called spin-spin relaxation and its relaxation rate is expressed by the inverse of the spin-spin relaxation time T_2. However, these seems to be misleading, because the transverse relaxation is induced not only by electron spin-electron spin interaction, and the observed kinetics of the relaxation cannot always be expressed by a single exponential function.

The transverse relaxation in solids is generally much faster than the longitudinal relaxation, because the transverse relaxation is very sensitive to the fluctuation of the precession frequency. The phase relaxation is often used synonymously for the transverse relaxation which, strictly speaking, includes both the free induction decay and the transverse relaxation of one spin packet. We shall use the phase relaxation exclusively to express the transverse relaxation of one spin packet.

The types of interaction causing the transverse relaxation are summarized also in Table 1.

2.4 Electron Spin Echo Spectrometer and Measurements

Our ESE spectrometer was first built in 1982, and since then has been continuously modified for improvement. The basic construction and characteristics of the spectrometer have already been reported in detail elsewhere [14, 15]. The present status will be briefly described here.

The spectrometer is made up from the combination of an X-band ESR spectrometer (JEOL, JES-RE1X) and a set of accessories for generating microwave pulses and registering ESE signals. Continuous microwave radiation from the ESR krystron is shaped into pulses of the shortest possible width of 30 ns by a PIN diode switch, amplified by a travelling-wave tube amplifier (Litton, Model 624) up to the highest possible power of 1 kW, and then fed into a home-made TE102 resonator or a loop-gap resonator through the ESR circulator. The resonators are designed to get a high rotating magnetic field and low Q. The microwave signals of the ESE are amplified by a GaAs-FET amplifier and detected with a homodyne detection system. The microwave pulses as well as a gate pulse for the ESE detection are controlled by a home-made pulse sequencer, which in turn is controlled by a personal computer. The digitized ESE signals are stored in the computer through a sample/hold circuit and an A/D converter. The ESE measurements are usually made by the 90°–180° two pulse mode. The time resolution of the total system is 2.5 ns, although the detection system becomes blind for 200 ns after each microwave pulse.

The phase relaxation curve was recorded by monitoring the peak intensity of the two-pulse ESE signal as a function of the time interval between the two microwave pulses, τ. The relative shape of an ESE signal is determined by the free induction decay, so that it does not depend on τ. The phase relaxation takes place between the time of the excitation and the observation, so that $t_2 = 2\tau$ is the time for the phase relaxation. The longitudinal relaxation also causes the decrease in the ESE intensity. However this effect is negligible because the rate of the longitudinal relaxation in solids is usually more than two orders of magnitude lower than the phase relaxation rate.

The longitudinal relaxation curve was recorded by either one of the two methods. In the first methods, the ESE intensity of the two-pulse mode with a fixed τ is monitored as a function of the repetition period t_1 of the ESE measurement. Since the ESE intensity is proportional to the amount of magnetization which is recovered from the xy-plane to the z-axis during the repetition period, the ESE intensity thus measured gives the longitudinal relaxation curve. In the second method, instead of changing the repetition period, a 90° pulse is applied at time t_1 before the two-pulse sequence. These are essentially the same as the ESR saturation recovery method, though the signal is monitored by the ESE technique.

2.5 Relaxation-Resolved ESR Detected by the Spin-Echo Method

Although the entire ESR spectrum of radical species in solids cannot be obtained by the Fourier transform of the ESE signal because of a large spectral width, it can be obtained by monitoring the ESE intensity while sweeping the external magnetic field very slowly. The shape of the ESR spectrum thus obtained is not necessarily the same as that of the cw ESR spectrum. The ESE method detects only the radical species which is not phase-relaxed at the time of detection, so that the radical species with very fast phase relaxation rate cannot be observed by the ESE-detected ESR.

A typical example is seen for 1-hydroxyethyl radical trapped in a γ-irradiated ethanol matrix at 77 K [16]. As is shown in Fig. 2, the cw ESR spectrum of the radical is composed of five lines due to hyperfine interactions with one α proton and three β protons of a methyl group. The hyperfine interaction depends on the location of the β protons with respect to the p_z orbital of the unpaired electron. However, the observed hyperfine coupling constant is the same for all the β protons because of the quick rotation of the methyl group in the time scale of the cw ESR measurement. On the other hand, the ESE-detected ESR spectrum is composed of four lines due to the hyperfine interactions with

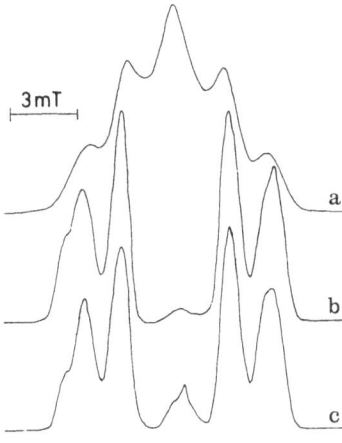

3 mT

a

b

c

Fig. 2a–c. ESR spectra of 1-hydroxyethyl radical generated by γ-irradiation of glassy ethanol at 77 K recorded
(**a**) by the usual cw-ESR method (displayed in the integ-
rated form), and by the t_1-resolved ESR method by using
the ESE technique (**b**) at $t_1 = 0$ μs and (**c**) at $t_1 = 10$ μs
with fixed t_2 of 1 μs

three β protons of the spin state $\pm (1/2, 1/2, 1/2)$ and one α proton of the spin
state $\pm 1/2$. The rotation of the methyl group with the nuclear spin state other
than $\pm (1/2, 1/2, 1/2)$ causes the fluctuation of the hyperfine interactions. The
rotation is not fast enough to average out the fluctuation of the hyperfine
interaction but is fast enough to induce the complete phase relaxation during the
time between the excitation and the detection. The 1-hydroxyethyl radical with
the methyl protons of the spin state other than $\pm (1/2, 1/2, 1/2)$ is therefore not
observed by the ESE-detected ESR.

Since the ESE-detected ESR measures an ESR spectrum as a function of t_2
and t_1, we can observe the ESR spectrum as a function of the longitudinal and
the phase relaxation degrees. We call this method relaxation-resolved ESR. The
relaxation resolved ESR can be used for differentiation of overlapped ESR
spectra from each other, when the relaxation rates of the coexisting radical
species are not the same. This provides us a powerful means for investigating the
nature and behavior of radicals in polymers, as described in Sect. 4.

2.6 Cross Relaxation and Spatial Distribution of Radicals

As described in Sects. 2.2 and 2.3, the inter-radical distance or the spatial
distribution of radical species in solids affects both the longitudinal relaxation
and the phase relaxation through electron spin-electron spin interactions.
Among them, the longitudinal relaxation is affected by the cross relaxation
mechanism, of which physical meaning is clear and the analysis based on the
spatial distribution of electron spins is straightforward. Although the phase
relaxation due to the instantaneous diffusion also provides us with a means of
determining the spatial distribution, the detection of the instantaneous diffusion
is experimentally difficult when the ESR spectrum of radical species is much
wider than the spectrum of A spins, as is generally the case for radical species in
polymers.

The cross relaxation is the energy transfer between A and B spins through dipolar interactions. Let us consider a pair of electron spins (an A spin and a B spin), the rate of the cross relaxation between them is given by [17, 18]

$$w = (\pi/12)S_B(S_B + 1)\gamma_A^2\gamma_B^2\hbar^2 r^{-6}(1 - 3\cos^2\theta)^2 J(\omega_B - \omega_A), \quad (1)$$

where S_B is the quantum number of the B spin, γ_A and γ_B are the gyromagnetic ratio of the A and the B spins, respectively, r is the distance between the A and the B spins, θ is the angle between the external magnetic field and the vector $\hat{r}$ joining the A and the B spins, and $J(\omega_B - \omega_A)$ is the spectral overlap function between the B and the A spins with the precession frequencies ω_B and ω_A, respectively.

When the A spin is surrounded by many B spins with different r and θ, the cross relaxation rate is given by the superposition of the effect from all the B spins with different r and θ, as

$$k = \sum_{i=1}^{N} w_i = \sum_{i=1}^{N} r_i^{-6}(1 - 3\cos^2\theta_i)^2 F_i(\omega_{B,i} - \omega_A), \quad (2)$$

where r_i, θ_i and $\omega_{B,i}$ are the distance, the angle and the frequency of the i-th B spin, respectively. The recovery of the longitudinal magnetization at time t_1 after the excitation follows the functional form of

$$V(t_1) = 1 - \exp(-kt_1). \quad (3)$$

The location of the B spins with respect to the A spin is actually not the same for all the A spins. Moreover, the precession frequency of the B spins is not all the same. The observed relaxation kinetics are therefore given by the superposition of Eq. (3) with a different rate constant.

When the electron spins are coupled with nuclear spins, the cross relaxation accompanying the change of the nuclear spin state can occur. In this case the apparent spectral overlap of the A and the B spins is not necessary. The spectral averaging of Eq. (3) is therefore a difficult task. Instead, we assume that the spectral overlap function in Eq. (2) is given by a constant F. Then, the spatial averaging of Eq. (3) is necessary for correlating the observed relaxation kinetics with the theory. The result of the spatial averaging will be shown for the two extreme cases of the spatial distribution of radicals in solids.

One of the extreme cases is a random distribution in which an A spin is surrounded by statistically-distributed B spins. This is the distribution for homogeneously-dispersed radicals. Summation of Eq. (3) under the random distribution gives

$$V(t_1) = 1 - \exp[-(16\pi^{3/2}/3^{5/2})C(Ft_1)^{1/2}] \quad (4)$$

where C is the number density of the radicals. Equation (4) shows that the relaxation rate is proportional to the concentration of the radicals and that the exponent of t_1 is 1/2. The parameter F can be determined by plotting the relaxation rate as a function of C. Taking the relaxation by the other processes,

$D(t_1)$, into account, the overall recovery kinetics is given by

$$V(t_1)/V_{max} = 1 - \exp[-(16\pi^{3/2}/3^{5/2})C(Ft_1)^{1/2}]D(t_1), \qquad (5)$$

where V_{max} is the intensity of the ESE signal without saturation.

Another extreme case is isolated radical pairs. Naturally radical species are generated by radiations pairwisely from diamagnetic molecules. This pairwise correlation will be retained in low-temperature solids. In a sample of very low radical concentration, the B spin in the pair is the only spin which opens an efficient path for the cross relaxation. The recovery kinetics in such a case is given by

$$V(t_1) = 1 - \int_0^\infty dr \int_0^\pi 2\pi r^2 \phi(r) \exp[-Fr^{-6}t_1(1 - 3\cos^2\theta)^2]\sin\theta\,d\theta, \quad (6)$$

where $\phi(r)$ is the distribution function of the intra-pair separation distance. The functional shape of the recovery kinetics depends on the distribution function and is not necessarily exponential.

A typical shape of the cross relaxation kinetics is an exponential function of the square-root of t_1. A particular distribution function

$$\phi(r) = 3\exp(-r^6/r_0^6)/(2\pi^{3/2}r_0^3), \qquad (7)$$

approximately gives this kind of recovery kinetics [19]:

$$V(t_1)/V_{max} = 1 - \exp[-1.04^3 r_0^{-3}(Ft_1)^{1/2}]D(t_1). \qquad (8)$$

Equations (5) and (8) shows that the exponential kinetics of the square-root of t_1 can be obtained for either the random distribution model or the radical-pair model. The same kinetic curve will be obtained from the number density of C_0 in the random distribution model and the characteristic distance r_0 in the radical-pair model, if

$$r_0 = 1.04[3^{5/2}/(16\pi^{3/2})]^{1/3}C_0^{-1/3}. \qquad (9)$$

Using the molar concentration instead of the number density,

$$r_0 = 0.689[R\cdot]_0^{-1/3}, \qquad (10)$$

where r_0 is in nm and $[R\cdot]_0$ is the concentration of the radical in mol/dm^3.

Under the distribution function of Eq. (7), the probability of finding a radical within radius r is 50% for $r = 0.781r_0$, 90% for $r = 1.051r_0$ and 99% for $r = 1.222r_0$. The γ-irradiation of solids generates isolated spurs composed of mainly one radical pair per spur [20]. If the distribution function of the intra-pair separation distance in the spur is given by Eq. (7), the average intra-pair separation distance, $\langle r\rangle$, is given by

$$\langle r\rangle = 0.781r_0. \qquad (11)$$

Defining the radius of the spur as the radius within which 99% of the radical is included, the spur radius r_s is expressed as

$$r_s = 1.565\langle r\rangle. \qquad (12)$$

The distribution function of Eq. (7) will be used for determining the structure of spurs in γ-irradiated polymers.

3 Free Radicals in Irradiated Polyethylene

3.1 Radiation Effects on Polyethylene

Polyethylene is known as one of the typical polymers which crosslink under the influence of ionizing radiation. The radiation-induced crosslinking (the formation of C–C covalent bonds between polymer chains) of polyethylene has long attracted the interest of a large number of research workers, because this polymer has the most simple chemical structure for the fundamental study of radiation effects on polymers, and also because the irradiation with ionizing radiation is a practically important means of modifying the mechanical and thermal properties of polyethylene.

The significant chemical effects of ionizing radiation on polyethylene are evolution of H_2, crosslinking between neighboring polymer chains, formation of transvinylene unsaturation, and destruction of crystallinity among many others [21]. The radiation yields of these effects depend on the polyethylene used, irradiation temperature, environment (in vacuum or in air), and other experimental conditions. Roughly, the G values at room temperature are 4, 2, and 1.5–1.8 for H_2 evolution, crosslinking and unsaturation formation, respectively. They are rather constant over a wide temperature range from 77 K, and they tend to increase above ca. 220 K [22].

Another important effect of radiation on polyethylene (generally on polymers) is the generation of free radicals. Three kinds of free radicals have been detected by the ESR method from polyethylene irradiated in the absence of oxygen: alkyl radical, $-CH_2\dot{C}HCH_2-$, allylic-type radical, $-CH_2CH=CH\dot{C}HCH_2-$, and polyenyl radical, $-CH_2-\dot{C}H-(-CH=CH-)_n-CH_2-$ in the order of increasing stability. For example, these radicals can be selectively observed by increasing the radiation dose typically to 10, 100, and 1,000 kGy, with γ-irradiation at room temperature [23]. The alkyl radical is exclusively generated by irradiating polyethylene under vacuum at low temperature (typically at 77 K) at a low dose, which exhibits an ESR spectrum with six hyperfine lines due to five protons on the α- and β-positions [23].

The recombination between two alkyl radicals is believed to be the main source of radiation-induced crosslinks in polyethylene [21], so that it is important to study the mechanism for formation and reaction of the alkyl radical. We have applied the ESE method to elucidate the paramagnetic relaxation mechanism and the spatial distribution of the alkyl radical, in order to get further insight into the radiation-chemical reactions of polyethylene resulting in the formation of crosslinks.

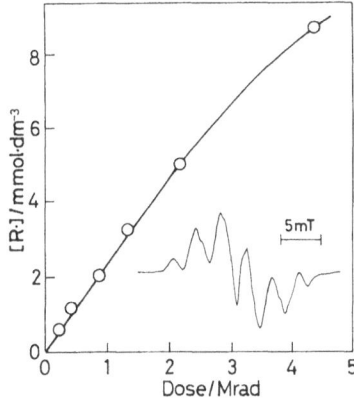

Fig. 3. Dependence of the concentration of the alkyl radical in polyethylene irradiated by γ-rays at 77 K in vacuum on the radiation dose (1 Mrad = 10 kGy). The *insert* shows the cw-ESR spectrum due to the alkyl radical

Figure 3 shows the dose-yield relationship for the formation of the alkyl radical at 77 K, together with the ESR spectrum in the insert, we observed for polyethylene sample of $M_v = 2.3 \times 10^4$ and density = 0.963 (CX-36F from the Showa Denko Co.). The slope of the initial linear portion gives the G value of 3.3 in accordance with the values reported previously [21, 24]. Above 30 kGy, the slope of the dose-yield curve gradually declines probably because of the overlapping of spurs.

3.2 Paramagnetic Relaxation of the Alkyl Radical

The alkyl radical in irradiated polyethylene was examined by the ESE technique to study its paramagnetic relaxation mechanism. Alkyl radicals in solid *n*-alkanes (C_6, C_8, C_{10}, and C_{12}) and cyclohexane irradiated at 77 K were also examined for comparison [25, 26]. Three kinds of alkyl radicals can possibly be generated in *n*-alkanes: chain end radicals $\dot{C}H_2CH_2-$, penultimate radicals $CH_3\dot{C}HCH_2-$, and inner radicals $-CH_2\dot{C}HCH_2-$. However, the chain end radicals are known to transform readily into the penultimate radicals [27]. Roughly speaking, the penultimate radicals give a seven-line ESR spectrum due to the hyperfine coupling of six protons, whereas the inner radicals give a six-line spectrum due to the hyperfine coupling of five protons. In cyclohexane, only the inner radicals can be generated owing to its cyclic structure; this situation is essentially the same as that for polyethylene with effectively infinite methylene chain.

Figure 4 shows the t_2-resolved ESR spectra of alkyl radicals in low-molecular weight alkanes and polyethylene obtained by the ESE technique described in Sect. 2. For cyclohexane and polyethylene, the spectral shape with six hyperfine lines is almost unchanged during the phase relaxation, consistent with the existence of only one kind of radical, the inner radical. For *n*-alkanes, contrarily, a weak seven-line spectrum due to the penultimate radical overlaps the six-line spectrum due to the inner radical at the early period of the phase relaxation (for

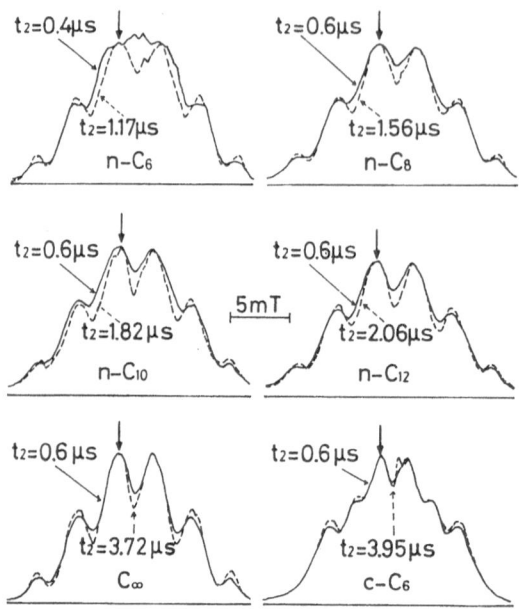

Fig. 4. The t_2-resolved ESR spectra, recorded at 77 K, of alkyl radicals in polyethylene, cyclohexane, and n-alkanes irradiated by γ-rays at 77 K to a dose of 7 kGy. The spectra of *broken lines* are magnified by a factor of 4 for ease of the comparison of spectral shapes. The *arrows* indicate the magnetic field position where the phase relaxation kinetics (Fig. 5) were examined

short t_2). The seven-line component decreases in intensity more rapidly than the six-line component. The phase relaxation of the penultimate radical is much faster probably because of the fluctuation of the hyperfine field due to the motion of the chain-end methyl group.

Figure 5 shows the change of the ESE intensity monitored at a peak of the six-line spectrum (indicated by arrows in Fig. 4) as a function of t_2. These phase relaxation curves do not follow a simple exponential function, but they are empirically expressed by an exponential of the 1.25th power of t_2. The jaggedness of the relaxation curves is due to the partially-resolved nuclear modulation effect of protons. The phase relaxation rate for n-alkanes given by the slope of

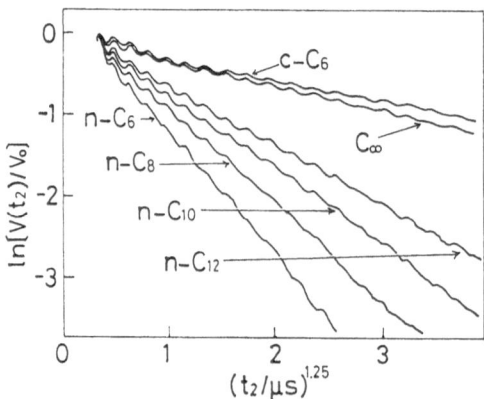

Fig. 5. Phase relaxation curves for the alkyl radicals in γ-irradiated polyethylene, cyclohexane, and n-alkanes. Signal intensities are normalized at $t_2 = 0.4$ μs

the relaxation curves is independent of the radical concentration and it decreases with an increase of the number of carbon atoms. In contrast, the phase relaxation rate for cyclohexane and polyethylene is considerably low and depends on the radical concentration. These findings are interpreted as due to two competing mechanisms for the phase relaxation in addition to a common mechanism due to the fluctuation of super-hyperfine interactions with the nuclear spins of surrounding diamagnetic molecules: the relaxation caused by the thermal fluctuation of hyperfine interaction (independent of the radical concentration) and the relaxation caused by the electron spin-electron spin interaction. The former mechanism is not operative in cyclohexane and polyethylene because of the molecular rigidity, so that the latter mechanism is predominant.

Inspection of the spectral shapes in Fig. 4 indicates that the linewidth of the six-line component due to the inner radical becomes sharpened with t_2. This spectral sharpening is interpreted as being due to the dependence of the relaxation rate on the resonance frequency: the relaxation is slower for the radicals at or near the peak of spectral lines than that for the radicals at the spectral tail because of the angular dependence of the hyperfine coupling.

Figure 6 shows the t_1-resolved ESR spectra of the alkyl radicals in alkanes and polyethylene obtained by using the ESE technique as described in Sect. 2.5. For cyclohexane and polyethylene, the spectral shape is essentially due only to the six-line component of the inner radical. For n-alkanes, the spectral shape is the superposition of the six-line component and the seven-line component. The latter due to the penultimate radical decays much more rapidly during the

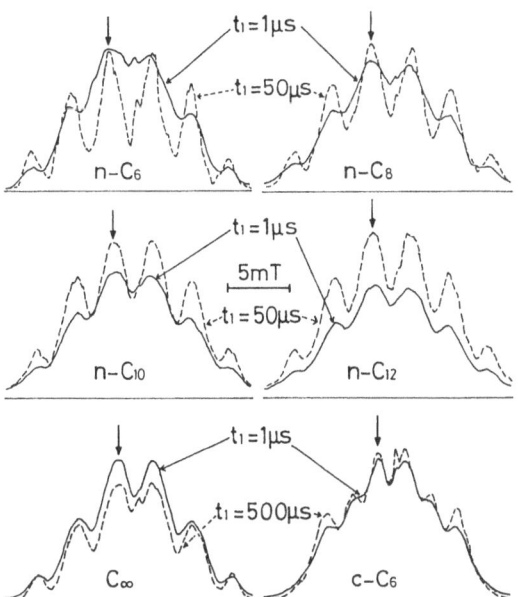

Fig. 6. The t_1-resolved ESR spectra with fixed t_2 of 0.6 μs, recorded at 77 K, of the alkyl radicals in polyethylene, cyclohexane, and n-alkanes irradiated by γ-rays at 77 K to a dose of 7 kGy. The spectra of *broken lines* are magnified by a factor of 4 for ease of the comparison of spectral shapes. The *arrows* indicate the magnetic field position where the longitudinal relaxation kinetics (Fig. 7) were examined

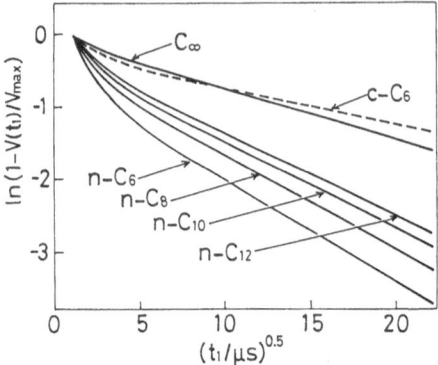

Fig. 7. Longitudinal relaxation curves for the alkyl radicals in γ-irradiated polyethylene, cyclohexane, and n-alkanes. Signal intensities are normalized at $t_1 = 1.0\,\mu s$

longitudinal relaxation. These features are essentially the same as those of the phase relaxation.

The longitudinal relaxation curves for the alkyl radicals monitored at the peak of the six-line spectrum (indicated by arrows in Fig. 6) are shown in Fig. 7. For n-alkanes, the relaxation rate decreases with the increase of the length of carbon chain, and it is independent of the radical concentration. This indicates that the cross relaxation between on-resonant and off-resonant radical spins is not the main source of the relaxation. In contrast, the relaxation rate for cyclohexane and polyethylene is the lowest, and depends on the radical concentration: it decreases with the decrease of the radical concentration. This indicates that the cross relaxation contributes to the longitudinal relaxation in cyclohexane and polyethylene.

In order to elucidate the mechanism for the rapid longitudinal relaxation of the inner radicals in n-alkanes, the ESE study was extended in detail to the alkyl radical in a n-hexane single crystal. The relaxation was found to be independent of the orientation of the crystal with respect to the external magnetic field. This shows that the relaxation is caused by the spectral diffusion induced by the modulation of isotropic hyperfine coupling. Analysis of the thermal motion of the radical reveals that the slow twisting of the C–C bond induces the modulation of the hyperfine coupling of β-protons [26]. In cyclohexane and polyethylene, this rapid longitudinal relaxation is not operative because of the molecular rigidity, so that the concentration-dependent cross relaxation becomes predominant.

3.3 Spatial Distribution of the Alkyl Radical

Both the phase and longitudinal relaxations for the alkyl radical in polyethylene are caused by the radical spin-spin interactions and are dependent on the radical concentration. This concentration dependence was examined in detail by changing the concentration with radiation dose and with thermal treatment [28]. The radiation dose was limited up to 30 kGy, where the G value of the alkyl radical is constant at 3.3.

An example of the phase relaxation curve observed for the alkyl radical in irradiated polyethylene has already been shown in Fig. 5. Regardless of the radiation dose, the relaxation kinetics monitored at the center of the ESR spectrum can be expressed empirically by the following equation:

$$V(t_2) = V_0 \exp(-\alpha t_2 - m t_2^2). \tag{13}$$

The rate constant for the first-order term, α, is dependent on the radiation dose (therefore, on the radical concentraion), while the rate constant for the second-order term, m, is independent of radiation dose. The dose-independent term is attributed to the hyperfine interaction with the surrounding protons.

The dose-dependent term of the phase relaxation is attributed to the electron spin-spin interaction as described in the preceding section. By taking the dose-yield curve shown in Fig. 3 into account, the dose-dependent rate constant can be empirically expressed as

$$\alpha = 0.275 + 0.045[R \cdot] \tag{14}$$

where α is in μs and $[R \cdot]$ is the bulk concentration of the radical in mmol/dm^3 determined by the cw ESR method.

The longitudinal relaxation of the alkyl radical in polyethylene as demonstrated in Fig. 7 can be expressed by the following equation:

$$V(t_1) = V_{max}\{1 - \exp(-b t_1^n)\}. \tag{15}$$

The exponent, n, is equal to 0.62 and is unchanged for all the radical concentrations examined. The rate constant, b, depends on the radiation dose and the thermal annealing after the irradiation.

The dependence of b on the bulk concentration of the radicals is shown in Fig. 8. The rate constant has a large intersect at zero-concentration and increases linearly with the bulk concentration as

$$b = 0.304 + 0.0284[R \cdot] \tag{16}$$

where b is in $(ms)^{-0.62}$. When the concentration is lowered by annealing an

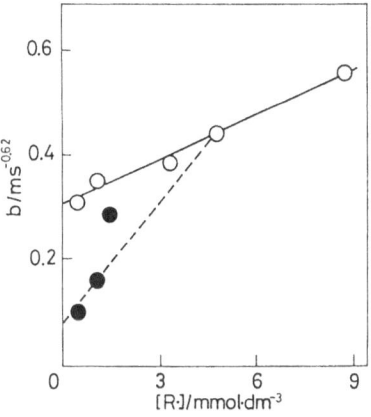

Fig. 8. The variation of the rate constant, b, of the longitudinal relaxation at 77 K of the alkyl radical in irradiated polyethylene with the change in the bulk radical concentration, ($\bigcirc$) during the concentration increase caused by the increasing radiation dose and ($\bullet$) during the concentration decrease caused by the successive thermal annealing

irradiated sample successively at 292, 313, and 323 K for a few minutes, the rate constant decreases much quicker and reaches a considerably smaller intersect. The thermal annealing does not cause the change of n. The value of the smaller intersect is regarded as the longitudinal relaxation rate for the final surviving radical, infinitely isolated from other radicals. Therefore, this is the relaxation intrinsic to the radical determined by the intramolecular mechanism, as elucidated for the alkyl radicals in the n-alkanes.

The linearly-increasing part of relaxation is attributed to the relaxation mechanism which prevails with the increase of the bulk concentration of the radical. It is due to the interaction with the radicals in other spurs randomly distributed around the radical under observation. The larger intersect corresponds to the relaxation for the radicals in the very first spur generated at the beginning of irradiation, so that it is the superposition of the relaxation rate for the intramolecular mechanism and that for the radical spin-spin interaction within the spur. Therefore, the difference between the larger and smaller intersects gives the relaxation rate solely due to the radical spin-spin interaction within a spur. The radicals are generated not in isolation but in a group (or in a pair) forming a spur. The effective local concentration of the radical in a spur can be estimated from the difference of intersects divided by the slope of the linear increase of the rate constant. The result shown in Fig. 8 gives the effective local concentration of 8.4 mmol/dm^3.

Assuming the radical pair model that each spur contains one pair of radicals, which has been confirmed as valid for spurs in the γ-irradiated glassy matrix of ethanol, and applying the theory of inter-radical spin-energy transfer due to the dipole-dipole interaction described in Sect. 2.5, [19], the average separation distance between the counterpart radicals in the pair is calculated to be 2.6 nm from the effective local concentration of 8.4 mmol/dm^3 and the distribution function given by Eq. (7). This means that the relaxation due to the counterpart radical of the average separation distance of 2.6 nm is equivalent to the relaxation due to the radicals randomly distributed with the concentration of 8.4 mmol/dm^3. Defining the spur radius as the radius within which 99% of the radical is included, the spur radius is estimated from the effective local concentration to be 4.1 nm.

The effective local concentration can be determined in the same way from the intersect and the slope of linear increase of the rate constant, α, of the phase relaxation as given by Eq. (14). It is calculated to be 6.2 mmol/dm^3, which is close enough to the value determined from the longitudinal relaxation.

The effective local concentration determined from the longitudinal relaxation as mentioned above is roughly equal to the value of the bulk concentration where the dose-yield curve begins to deviate from the linear relationship. Assuming that the dose-yield curve starts to saturate at the bulk concentration of 8.4 mmol/dm^3 or the radical-pair concentration of 4.2 mmol/dm^3 due to the overlap of the spurs, the radius of the spur is estimated to be 4.5 nm. This value is in good agreement with the spur radius of 4.1 nm obtained from the relaxation measurement. This coincidence seems to support the general view that the decrease of the G value for the radical formation at high radiation doses is caused by spatial overlapping of spurs.

The theoretical treatment to correlate the separation distance and the local concentration presumes that the relaxation is expressed by the exponential function of $t_1^{1/2}$. The observed exponent of t_1 is 0.62 and deviates from the theoretically expected value of 0.5. This difference is attributed to the effect of the energy transfer within the radicals on resonance (between A spins), which has not been taken into account in the theory developed in Sect. 2.6. Nevertheless, the heterogeneous distribution has been unequivocally demonstrated by the ESE method.

3.4 Reaction Mechanism for Alkyl Radical Formation

The primary event caused by radiation is ionization or direct excitation of polyethylene chains, RH, which leads to the alkyl radical formation. Toriyama et al. [4] proposed a reaction mechanism for the radiation effect on polyethylene based on the ESR study. They presumed that an alkyl radical and an H atom were generated either directly from the dissociation of the excited state polymer or from the recombination between the cation radical and the electron primarily generated by ionization. There are three possible reaction paths involving the charge recombination:

$$RH\cdot^+ + e^- \longrightarrow RH* \tag{17}$$

$$RH* \longrightarrow R\cdot + H\cdot \tag{18}$$

$$RH\cdot^+ + RH \longrightarrow R\cdot + RH_2^+ \tag{19}$$

$$RH_2^+ + e^- \longrightarrow RH + H\cdot \tag{20}$$

$$RH\cdot^+ \longrightarrow R^+ + H\cdot \tag{21}$$

$$R^+ + e^- \longrightarrow R\cdot \tag{22}$$

$$H\cdot + RH \longrightarrow R\cdot + H_2. \tag{23}$$

The first reaction path, (17) and (18), proceeds through the charge recombination between the cation radical and the electron, the second path, (19) and (20), proceeds through the proton transfer from the radical cation, and the third path, (21) and (22), proceed through the H atom release from the radical cation. Although it has not been known which is the most dominating among the competing reactions, (17), (19), and (21), of the radical cation, the latter two are thought to be important [4]. In any case, the H atom is generated with an alkyl radical and it reacts with a nearby polyethylene chain to form another alkyl radical (reaction (23)), so that two alkyl radicals are generated in pair. Recently, Brede et al. have shown by the pulse radiolysis of molten polyethylene that the dissociation of C–H bond occurs very rapidly probably through the excitation migration [29]. In this reaction mechanism also, the alkyl radicals are generated in pair. The separation distance between the two alkyl radicals in the pair is determined by the migration of the H atom before reacting with a nearby polyethylene chain. It is as long as 2.6 nm on average.

Early in 1961, Libby [30] suggested the proton transfer from the radical cation and subsequent charge recombination between the protonated site of polyethylene chain and an electron for the alkyl radical formation:

$$RH^{.+} + RH \longrightarrow R^. + RH_2^+ \tag{19}$$

$$RH_2^+ + e^- \longrightarrow R^. + H_2 \tag{24}$$

This reaction mechanism predicts a small separation distance between the two alkyl radicals in pair determined by the proton-transfer distance. This small separation distance was thought to favor the high yield of the radiation-induced crosslinks. However, the above reaction mechanism seems not to accommodate the results of the ESE study.

4 Free Radicals in Irradiated Poly(methly methacrylate)

4.1 Radiation Chemistry of Poly(methyl methacrylate)

Since the early days of radiation-chemical studies on polymers, poly(methyl methacrylate) (abbreviated as PMMA in this chapter) has been known to be a typical radiation-degradable polymer [22]. Polymers are roughly divided into two classes: radiation-crosslinking polymers and radiation-degradable polymers. The polymers having the chemical structure of $-CH_2-CR^1R^2-$ usually exhibit the radiation-degrading nature. PMMA with $R^1 = -CH_3$ and $R^2 = -COOCH_3$ belongs to this class. The G value of 2.6 for the scission of main chains seems to be a standard value, while the G value for the crosslinking is practically zero [22]. Owing to this purely degrading nature, PMMA is now regarded as a promising positive resist for X-ray and electron-beam lithography. The elucidation of the reaction mechanism of the radiation-induced degradation of PMMA has long been and is still one of the central themes in radiation chemistry of polymers from both the fundamental and practical viewpoints.

Among numerous previous studies on the reaction mechanism, Geuskens et al. [31] proposed the most detailed mechanism based on the analysis of radiation products as well as the effect of added ethylmercaptan:

$$\underset{\substack{| \\ O=COCH_3}}{\overset{\substack{CH_3 \\ |}}{-CH_2-C-CH_2-}} \rightsquigarrow \underset{\substack{+}}{\overset{\substack{CH_3 \\ |}}{-CH_2-C-CH_2-}} + \;^.COOCH_3 + e^- \tag{25}$$

$$\underset{+}{-CH_2-\overset{\overset{\displaystyle CH_3}{|}}{\underset{|}{C}}-CH_2-} \ + \ e^- \ \longrightarrow \ -CH_2-\overset{\overset{\displaystyle CH_3 *}{|}}{\underset{\overset{\displaystyle |}{\bullet}}{C}}-CH_2- \tag{26}$$

$$-CH_2-\overset{\overset{\displaystyle CH_3 *}{|}}{\underset{\overset{\displaystyle |}{\bullet}}{C}}-CH_2- \ \longrightarrow \ -CH_2-\overset{\overset{\displaystyle CH_3}{|}}{\underset{|}{C}}=CH_2 \ + \ \overset{\overset{\displaystyle CH_3}{|}}{\underset{\overset{\displaystyle |}{O=COCH_3}}{\bullet C}}-CH_2- \tag{27}$$

They assumed that the primary cation radical of PMMA spontaneously and quickly dissociated to form carbocation, which then recombined with the liberated electron to form an excited radical with a *tert*-alkyl structure. This excited radical was thought to be the precursor of the scission of the main chain. This reaction model interpreted well their observation that the G value for the scission of the side chain was close to that of the main chain and that the mercaptan added to scavenge electrons suppressed the main-chain scission efficiently without affecting the formation of volatile products from the ester side chain. The above reaction model motivated us to apply the ESE method to the study of radicals in irradiated PMMA. The model now seems inadequate, because it cannot accommodate some recent ESE results as mentioned later.

PMMA has been known, since the 1950s, to show the ESR spectrum with five intense and four weak hyperfine lines (see the insert of Fig. 9), when it is irradiated with γ-rays at room temperature. After a long history of the study on this anomalous ESR spectrum [32–34], the interpretation is now almost settled that it is due to the propagating-type radical $-CH_2-\dot{C}(CH_3)COOCH_3$ (the radical expected to form during the radical polymerization of methyl methacrylate) [35, 36]. Although the formation of this radical is a definite proof of the radiation-induced scission of the PMMA main-chain, the previous ESR studies have failed to elucidate the mechanism for the formation of this radical.

One of the powerful approach for elucidating the mechanism of radical reactions is the irradiation of sample specimens with γ-rays at 77 K followed by

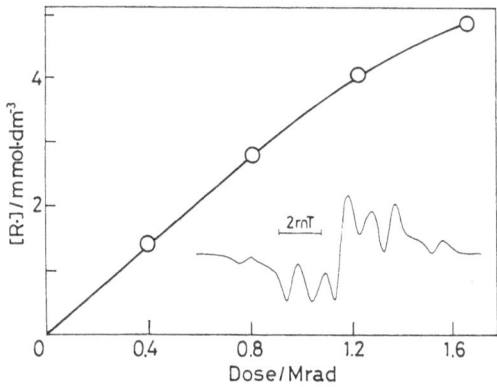

Fig. 9. Dependence of the concentration of the radicals in PMMA irradiated by γ-rays at 273 K in vacuum on the radiation dose (1 Mrad = 10 kGy). The insert shows the cw-ESR spectrum from which the concentration is determined

monitoring the ESR signal during subsequent warming. However, this approach has been difficult to be applied to PMMA. The change of the ESR spectral shape during warming is too complex, though the 5 + 4 line spectrum can generally be observed at room temperature. The coexistence of several radical species has made it impossible to separate each component spectrum from the others. This situation is different from that for polyethylene.

In order to get further insight into the reaction mechanism for the degradation of PMMA, we have studied the nature and behavior of radical entities in irradiated PMMA by using the ESR and ESE techniques complementarily [37]. Two PMMA samples, a commerical PMMA and an initiator-free PMMA prepared by the radiation-polymerization of bulk monomer, were used, but no difference was found in the results. Residual monomer was carefully removed from the PMMA samples, because the monomer molecule readily modifies the radicals derived from the polymer. The samples were irradiated in vaccum. Figure 9 demonstrates the dose-yield curve we obtained by irradiating PMMA in vacuum at 273 K. The G value for the radical formation is determined to be 3.0 from the slope of the linear portion below 12 kGy.

4.2 Identification of ESR Spectra

The ESR spectrum of PMMA recorded immediately after the γ-irradiation at 77 K is the superposition of several components with a different hyperfine structure, as shown in Fig. 10(A). It has been determined as being the superposition of a central broad singlet due to the anion radical with an electron localized on the C=O bond in the side group [38], a triple spectrum with a hyperfine coupling constant of 2 mT due to the side-chain radical, $-COO\dot{C}H_2$ [39], a sharp quartet spectrum with a coupling constant of 2.3 mT due to the methyl radical [39], and a widely spread doublet spectrum with a coupling constant of 13 mT due to the formyl radical, $\dot{C}HO$ [40]. Photobleaching of the anion radical with visible light enhances the spectrum of the methyl radical as is shown in Fig. 10(B).

The total G value for radical formation at 77 K was found to be 4.2. The G value for the photobleached anion radical is 1.2 in accordance with the previous report [41]. The photobleaching transforms the anion radical into the methyl radical with an efficiency of about 10%. A small quantity of the methyl radical exhibits a very prominent spectrum in the cw-ESR measurements because of the narrow width of the spectral lines.

The ESE method was used to identify the overlapping ESR spectra of the irradiated PMMA more definitely. The measurements of the ESE-detected ESR were made at 77 K with the $90° - \tau - 180°$ two-pulse sequence, at various fixed times of longitudinal relaxation, t_1, while the external magnetic field was swept slowly. The t_1-resolved ESR spectrum was obtained from the difference between the echo intensity at a fixed τ of 0.5 μs with and without a 90° saturation pulse

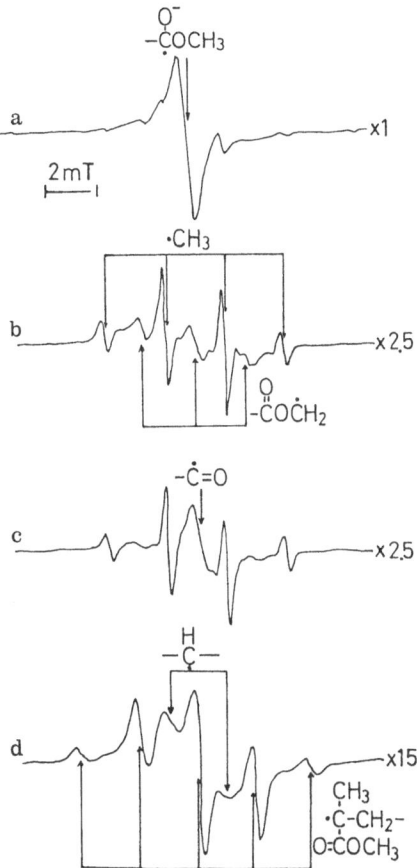

Fig. 10a–d. The cw-ESR spectra of PMMA γ-irradiated to a dose of 5 kGy and recorded both at 77 K (a) immediately after irradiation, (b) after bleaching with visible light, (c) after subsequent bleaching with UV light, and (d) after subsequent thermal annealing at room temperature

preceded by t_1. The microwave pulses of 7.5 W incident power with the widths of 120 ns and 240 ns were used for the 90° and 180° pulses in this study.

Figure 11 shows the t_1-resolved ESR spectra of the irradiated PMMA before and after the photobleaching at 77 K. The broad singlet line at the center of the spectrum in Fig. 11(A) is due to the anion radical. The anion radical signal becomes comparatively less intense with the increase of t_1, and a doublet signal with a coupling constant of 2 mT becomes apparent. This doublet can be attributed to the main-chain radical, $-\dot{C}H-$ [39, 42], and is evidenced for the first time to exist immediately after the irradiation by using the ESE method. The t_1-resolved ESR spectra after the photobleaching (Fig. 11(B)) show clearly the doublet due to the main-chain radical and the triplet due to the side-chain radical, because the spectrum of the methyl radical decays away very quickly due to its fast phase relaxation. On the other hand, the spectrum of the main-chain radical is not observed by the conventional cw ESR method, because it was masked by the strong spectral lines of the methyl radical.

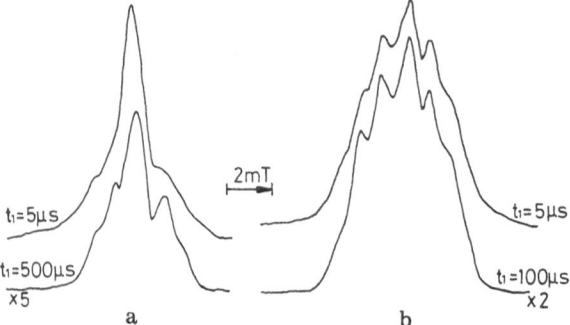

Fig. 11a, b. The t_1-resolved ESR spectra, recorded with fixed t_2 of 1.0 μs, of PMMA γ-irradiated to a dose of 5 kGy (**a**) before and (**b**) after photobleaching with visible light. All the irradiation, photobleaching and measurements were made at 77 K

Based on these ESR and ESE results, the G values for the initial formation of radicals at 77 K were determined to be 2.0, 1.2, and 1.0 for the side-chain radical, the anion radical, and the main-chain radical, respectively.

4.3 Thermal and Photo-Induced Reactions of Radicals

When the irradiated sample is progressively warmed to room temperature, the ESR, both cw and t_1-resolved, spectra change in a complex manner. The final shape of the spectra is the superposition of the two components: the two-line spectrum due to the main-chain radical and the 4 + 5 line spectrum due to the propagating-type radical. The coexistence of the main-chain radical with the propagating-type radical is definitely shown for the first time. The residual monomer, if present in the sample PMMA, readily adds to the main-chain radical, so that the ESR spectrum is changed to one with sharp and symmetric hyperfine structure [35] due to the propagating-type radical. The G values of the radicals remaining at room temperature are 0.8 and 2.0 for the main-chain radical and the propagating-type radical, respectively. The latter value is close to the G value for the scission of the PMMA main chain.

The spectral change during this progressive warming to room temperature is shown in Fig. 12 for the sample irradiated and photobleached at 77 K. Elimination of the anion radical by photobleaching with visible light makes the analysis of the spectral shape easy. Then the observed spectral change indicates:

(1) The methyl radical spectrum enhanced by the photobleaching decays at 77–180 K. The disappearance of the methyl radical is partly complemented by the formation of the side chain radical, $-COO\dot{C}H_2$, due to the abstraction of a hydrogen atom from the side chain ester group, $-COOCH_3$.

(2) At 230–265 K, the side-chain radical transforms into the propagating-type radical, $-CH_2-\dot{C}(CH_3)COOCH_3$. This conversion proceeds without loss

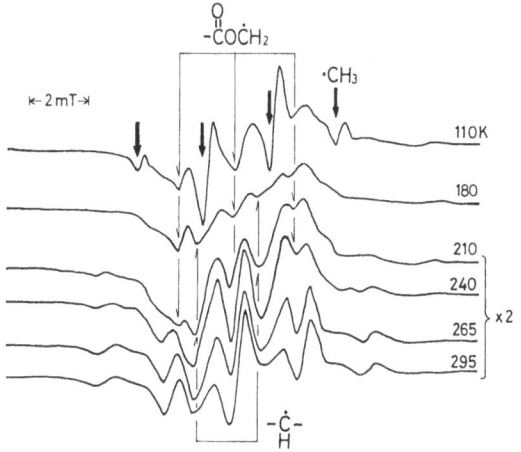

Fig. 12. Thermal change of the cw-ESR of PMMA γ-irradiated to a dose of 5 kGy and photobleached with visible light at 77 K. Each spectrum was recorded after successive warming to temperatures indicated in the figure for 15 min. The spectrum for 295 K was recorded exceptionally after thermostating for 24 hours

of total radical concentration, so that the G value of the propagating-type radical present at room temperature is 2.

(3) The main-chain radical, $-\dot{C}H-$, partly disappears above 265 K but most of the radical (G = 0.8) survives at room temperature. The decay of this radical at either below or above room temperatures does not accompany the formation of any new radical.

(4) If the sample is warmed without photobleaching, the anion radical decays at 77–180 K accompanying the partial transformation into the side-chain radical.

No difference in the spectral features was observed above 180 K between the samples with and without the elimination of the anion radical by photobleaching prior to warming.

Figure 10 shows the photo-induced transformation of radical species in irradiated PMMA [43]. The photobleaching with visible light at 77 K causes the liberation of an electron from the anion radical and the splitting of a methyl radical from the ester side group:

$$
\begin{array}{cccc}
& CH_3 & & CH_3 \\
& | & & | \\
-CH_2-C- & + h\nu(vis) \longrightarrow & -CH_2-C- & + \cdot CH_3 \\
& | & & | \\
& {}^{-}O-\dot{C}OCH_3 & & {}^{-}O-C=O
\end{array}
\qquad (28)
$$

The formation of ethyl radical from the similar photobleaching experiment for poly(ethyl methacrylate) definitely indicates that the methyl radical is not from the elimination of the α-methyl group, as once postulated [38], but is from the splitting of methyl group in the ester side chain.

Further irradiation with UV light causes the formation of new radical species with a broad singlet spectrum at the expenses of the triplet spectrum due to the side-chain radical, as demonstrated in Fig. 10(C) [43]. The new radical species is assigned to the acyl-type radical, $-\dot{C}=O$, generated by the loss of CH_2O from the side-chain radical. The photo-generation of the acyl-type radicals has been found in γ-irradiated low molecular weight carboxylic esters [44]:

$$R—COO\dot{C}HR' \longrightarrow R—\dot{C}{=}O + R'CHO \qquad (29)$$

The similar photo-induced radical conversion has been confirmed also for methyl pivalate, $C(CH_3)_3COOCH_3$, as a model compound of PMMA [43].

Heating the UV-irradiated sample to room temperature causes the quantitative conversion of the acyl-type radical into the propagating-type radical as is shown in Fig. 10(D). The shape and intensity of the ESR spectrum after heating the visible- and UV-irradiated sample were found to be identical with those after heating the visible light-irradiated sample without the conversion of the side-chain radical into the acyl-type radical. This strongly suggests that both the side-chain radical and the acyl-type radical are thermally converted to the same radical species, which are then spontaneously converted to the propagating-type radical. The most plausible direct precursor state for the scission of the main chain is the alkyl radical on the tertiary carbon atom, $-CH_2-\dot{C}(CH_3)-CH_2-$.

It should be noted that most of the anion radical simply disappears without giving any new radical species when photobleached with visible light. This observation seems to exclude reaction (26), the key step in the reaction mechanism proposed by Geuskens et al. [31]. The photo-induced disappearance of the anion radical must release an electron, which recombines with some cationic entities. This charge recombination does not actually lead to the formation of the propagating-type radical, in contradiction with the proposed reaction mechanism.

Very recently, Plaček and Szöcs have reported the ESR study of radicals generated in monomer-free PMMA at low temperatures by γ-irradiation They have shown the thermal behavior of several types of radicals above 100 K [45]. There is no essential difference of spectral assignment between their results and the present ones, except that they incorrectly assigned the singlet spectrum due to the anion radical as being due to acyl-type radicals.

4.4 Paramagnetic Relaxation and Spatial Distribution of Radicals

The total G value is about 3.0 for the radical formation in PMMA irradiated at room temperature. This value is very close to the G value of total radicals remaining at room temperature after the irradiation at 77 K. The ESR spectral shape observed from the PMMA irradiated at room temperature is the same as that observed from the PMMA irradiated at 77 K and warmed to room temperature. These results strongly suggest that the radiation-induced reactions in PMMA are independent of irradiation temperature. The concentration ratio of the main-chain radical to the propagating-type radical is 1/2. Paramagnetic relaxation of the free radicals in PMMA irradiated at room temperature was examined as a function of the radiation dose and of the bulk radical concentration, in the same way as for the alkyl radical in irradiated polyethylene.

The phase relaxation was found to follow a simple exponential function,

$$V(t_2) = V_0 \exp(-\alpha t_2) \tag{30}$$

with the rate constant, in μs^{-1} unit, expressed as

$$\alpha = 0.475 + 0.043[R \cdot], \tag{31}$$

where $[R \cdot]$ represents the bulk concentration of the total radicals in $mmol/dm^3$.

The longitudinal relaxation was found to be expressed as

$$V(t_1) = V_{max}\{1 - \exp(-bt_1^{0.62})\}. \tag{32}$$

The exponent of t_1, 0.62, is unchanged irrespective of the radiation dose and of the thermal annealing, and it is identical with the exponent determined for the alkyl radical in polyethylene. The rate constant, b, changes depending on the dose and the annealing treatment, as is shown in Fig. 13.

When the bulk radical concentration increases with the radiation dose, the rate constant increases linearly starting with a large intersect (open circles). The rate constant is expressed as a function of the radical concentration, $[R \cdot]$, as

$$b = 1.55 + 0.114[R \cdot], \tag{33}$$

where b is in $ms^{-0.62}$.

When the concentration is decreased by successive thermal annealing at 360 K, the rate constant decreases more quickly to give a smaller intersect (closed circles). The difference between the two intersects divided by the slope of

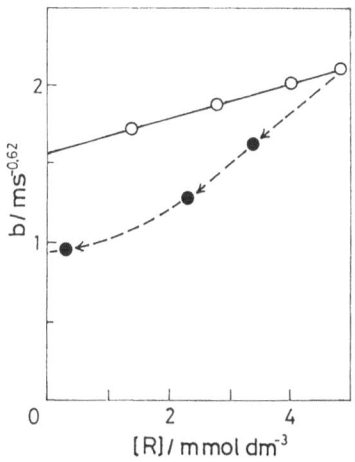

Fig. 13. The variation of the rate constant, b, of the longitudinal relaxation at 77 K of the radicals in PMMA γ-irradiated at 273 K with the change in the bulk radical concentration, (○) during the concentration increase caused by the increasing radiation dose and (●) during the concentration decrease caused by the successive thermal annealing

the linear increase gives the effective local concentration of the radicals, 5.4 mmol/dm^3. Using the same procedure as in Sect. 3.3, the local concentration of 5.4 mmol/dm^3 leads to the average intra-pair separation distance of 3.1 nm and the spur radius of 4.8 nm.

The effective local concentration of 5.4 mmol/dm^3 is roughly equal to the bulk concentration where the radical yield starts to saturate. Assuming that the spur starts to overlap at the bulk concentration of 5.4 mmol/dm^3 or the radical pair concentration of 2.7 mmol/dm^3, the radius of the spur containing a pair of radicals is estimated to be 5.3 nm. The spur radius obtained from the relaxation measurement is in good agreement with that obtained from the dose-yield curve, which implies that the decrease of the radical yield at high doses is really due to the overlapping of the spurs.

As mentioned in the previous section, the concentration of the neutral polymer radicals in PMMA is scarcely changed by thermal annealing at room temperature. The migration of the polymer radicals is absent and the initial intra-pair separation distance at 77 K is maintained at room temperature. The average intra-pair separation distance of 3.1 nm is much shorter than that of cation-anion pairs, 4.5–5 nm, [20] in γ-irradiated rigid organic glasses and is close to that of alkyl radical pairs, 2.6 nm, in γ-irradiated polyethylene. The cation-anion separation is primarily determined by the migration distance of electrons before being stabilized, whereas the alkyl radical intra-pair separation is determined by the migration distance of hydrogen atoms before abstracting a hydrogen atom from the C–H bond.

Geuskens et al. assumed that the counterpart of a polymer radical is a ĊOOCH$_3$ radical. However a singlet ESR line which they assigned to the ĊOOCH$_3$ radicals is actually due to the anion radicals. Therefore, for explaining the ESR results based on the Geuskens' mechanism, it is necessary to assume that the ĊOOCH$_3$ radical reacts with PMMA at 77 K to convert to a polymer radical. The intra-pair separation distance would be much shorter than 3.1 nm if the initial radical is ĊOOCH$_3$, because even the smaller radicals, ĊH$_3$ and ĊHO, cannot migrate at 77 K. The value of 3.1 nm for PMMA indicates that the initial counterpart of the polymer radical is not the large ĊOOCH$_3$ radical but is the small hydrogen atom.

4.5 Mechanism of Radiation-Induced Degradation

The propagating-type radical, a finger-print of the main-chain scission, is generated secondarily through the transformation of the side-chain radical. This indicates that the main-chain scission is preceded by the formation of the side-chain radical. The radical transformation occurs with the efficiency of practically unity at relatively high temperature where the polymer chains become mobile. Therefore, the key step for the main chain scission is the formation of the side-chain radical.

The primary active entities generated by the ionizing radiation are a cation radical, an excess electron, and an excited state PMMA:

$$
\begin{array}{ccc}
\overset{\displaystyle CH_3}{\underset{\displaystyle O=\overset{|}{C}OCH_3}{-CH_2-\overset{|}{\underset{|}{C}}-}}
& \rightsquigarrow \quad
\overset{\displaystyle CH_3{}^{\cdot\,+}}{\underset{\displaystyle O=\overset{|}{C}OCH_3}{-CH_2-\overset{|}{\underset{|}{C}}-}} + e^-,
& \overset{\displaystyle CH_3\,*}{\underset{\displaystyle O=\overset{|}{C}OCH_3}{-CH_2-\overset{|}{\underset{|}{C}}-}}
\end{array}
\tag{34}
$$

The electron is efficiently captured by the ester side-group to form the anion radical [46]. The side-chain and the main-chain radical is generated probably from the cation radical through intramolecular proton transfer or intermolecular proton transfer (as is known for small organic molecules [27, 47]):

$$
\overset{\displaystyle CH_3{}^{\cdot\,+}}{\underset{\displaystyle O=\overset{|}{C}OCH_3}{-CH_2-\overset{|}{\underset{|}{C}}-}}
\quad \longrightarrow \quad
\overset{\displaystyle CH_3}{\underset{\displaystyle {}^+HO-\overset{|}{C}O\dot{C}H_2}{-\overset{|}{\underset{|}{C}}H_2-C-}}
\tag{35}
$$

$$
\overset{\displaystyle CH_3{}^{\cdot\,+}}{\underset{\displaystyle O=\overset{|}{C}OCH_3}{-CH_2-\overset{|}{\underset{|}{C}}-}}
\quad \xrightarrow{\text{PMMA}} \quad
PMMA \cdots H^+ \; + \;
\overset{\displaystyle CH_3}{\underset{\displaystyle O=\overset{|}{C}OCH_3}{-CH-\overset{|}{\underset{|}{C}}-}} \;,
$$

$$
\overset{\displaystyle CH_3}{\underset{\displaystyle O=\overset{|}{C}O\dot{C}H_2}{-CH_2-\overset{|}{\underset{|}{C}}-}}
\tag{36}
$$

These intra- and inter-molecular deprotonations of the cation radical readily occur, so that the cation radical is unstable even at 4 K. The direct scission of a C–H bond in the excited state PMMA cannot be excluded as an additional source of either of the main-chain radical or the side-chain radical. The H atom generated from the excited state PMMA or from the recombination of a proton with an electron will recombine with the polymer radical or abstract an hydrogen to generate also the main-chain or the side-chain radical as

$$
\overset{\displaystyle CH_3}{\underset{\displaystyle O=\overset{|}{C}OCH_3}{-CH_2-\overset{|}{\underset{|}{C}}-}} + H\cdot \longrightarrow
\overset{\displaystyle CH_3}{\underset{\displaystyle O=\overset{|}{C}OCH_3}{-\dot{C}H-\overset{|}{\underset{|}{C}}-}} \;,\;
\overset{\displaystyle CH_3}{\underset{\displaystyle O=\overset{|}{C}O\dot{C}H_2}{-CH_2-\overset{|}{\underset{|}{C}}-}}
\tag{37}
$$

The intra-pair separation distance is therefore determined by the migration distance of the hydrogen atom before reacting.

The present study by the ESR and ESE methods confirms the transformation of the side-chain radical into the propagating-type radical. This radical transformation very plausibly proceeds as

$$
\begin{array}{ccc}
& \text{CH}_3 & \text{CH}_3 \\
& | & | \\
-\text{CH}_2-\text{C}- & \longrightarrow & -\text{CH}_2-\text{C}- \quad + \quad \text{CO, OCH}_2, \\
& | & | \\
& \text{O}=\text{CO}\dot{\text{C}}\text{H}_2 &
\end{array} \tag{38}
$$

and

$$
\begin{array}{cccc}
\text{CH}_3 & \text{CH}_3 & & \text{CH}_3 \\
| & | & & | \\
-\text{CH}_2-\underset{\cdot}{\text{C}}-\text{CH}_2-\text{C}-\text{CH}_2- & \longrightarrow & -\text{CH}_2-\text{C}=\text{CH}_2 \\
& | & & \\
& \text{O}=\text{CO}\text{CH}_3 & &
\end{array}
$$

$$
\begin{array}{cc}
& \text{CH}_3 \\
& | \\
+ \quad & \cdot\text{C}-\text{CH}_2- \\
& | \\
& \text{O}=\text{CO}\text{CH}_3
\end{array} \tag{39}
$$

The transient radical, $-\text{CH}_2-\dot{\text{C}}(\text{CH}_3)-\text{CH}_2-$, has not yet been observed. However, it has been shown that the evolution of CO and CH_2O gas are the main radiation-chemical products from PMMA [48, 49], and C=C double bond forms during the radiolysis of PMMA [50]. The effect of UV-irradiation mentioned previously also supports the reaction path of (38) and (39). Since the acyl-type radical is quantitatively converted to the propagating-type radical as the side-chain radical is, it is reasonable to assume that the above transient radical is the common intermediate for the two reactions: the radical conversions to the propagating-type radical from the side-chain radical and from the acyl-type radical. The reaction mechanism involving the acyl-type radical will be expressed as

$$
\begin{array}{ccc}
\text{CH}_3 & & \text{CH}_3 \\
| & & | \\
-\text{CH}_2-\text{C}- \quad + \quad \text{hv(UV)} & \longrightarrow & -\text{CH}_2\text{C}- \quad + \quad \text{CH}_2\text{O} \\
| & & | \\
\text{O}=\text{CO}\dot{\text{C}}\text{H}_2 & & \text{O}=\text{C}\cdot
\end{array} \tag{40}
$$

and

$$
\begin{array}{ccc}
\text{CH}_3 & \text{CH}_3 \quad \text{CH}_3 & \\
| & | \qquad | & \\
-\text{CH}_2-\text{C}- & \longrightarrow \quad -\text{CH}_2\underset{\cdot}{\text{C}}-\text{CH}_2-\text{C}-\text{CH}_2- \quad + \quad \text{CO} \\
| & | & \\
\text{O}=\text{C}\cdot & \text{C}=\text{OOCH}_3 &
\end{array} \tag{41}
$$

followed by reaction (39).

In summary, the side-chain radical of the structure $-COO\dot{C}H_2$ is the direct precursor and plays a key role in the radiation-induced scission of PMMA main-chain. Therefore, the main-chain scission can be suppressed by inhibiting the formation of the side-chain radical or by killing it with an adequate scavenger. On the contrary, the enhancement of the formation of the side-chain radical will be a guiding principle to increase the sensitivity of PMMA as an electron-beam resist.

5 References

1. Rånby B, Rabek JF (1977) ESR spectroscopy in polymer research. Springer, Berlin Heidelberg New York
2. Bullock AT, Sutcliffe LH (1964) Trans Faraday Soc 60: 211
3. Yoshida H, Hayashi K, Okamura S (1964) Ark Kemi 23: 177
4. Toriyama K, Muto H, Nunome K, Fukaya M, Iwasaki M (1981) Radiat Phys Chem 18: 1041
5. Shimada S, Hori Y, Kashiwabara H (1982) Radiat Phys Chem 19: 33
6. Mozumdar A, Magee JL (1966) Radiat Res 28: 203
7. Kowari S, Sato S (1978) Bull Chem Soc Jpn 51: 741
8. Kaplan DE (1962) Rev Sci Instr 31: 1182
9. Kevan L, Schwartz RN (eds) (1979) Time domain electron spin resonance. Wiley, New York
10. Mims WB (1972) Electron spin echo. In: Geschwind S (ed) Electron paramagnetic resonance. Plenum, New York, p 263
11. Trifunac AD, Norris JR, Lawler RG (1979) J Chem Phys 71: 4380
12. Abragam A (1961) The principle of nuclear magnetism. Oxford University Press, London
13. Kurshev VV, Raitsimring AM, Tsvetkov YD (1989) J Magn Reson 81: 441
14. Ichikawa T (1986) J Magn Reson 70: 280
15. Ichikawa T, Yoshida H, Westerling J (1989) J Magn Reson 85: 132
16. Ichikawa T, Yoshida H (1990) J Phys Chem 94: 949
17. Salikhov KM, Semenov SG, Tsvetkov YD (1976) Electron spin echo and its applications. (In Russ) Nauka, Novosibirsk
18. Bowman MK, Norris JR (1982) J Phys Chem 86: 3385
19. Ichikawa T, Yoshida H (1984) J Phys Chem 88: 3199
20. Ichikawa T, Wakasugi S, Yoshida H (1985) J Phys Chem 89: 3583
21. Dole M (1974) Radiation chemistry of polyethylene. In: Burton M, Magee JL (eds) Advances in radiation chemistry, vol 4. Wiley, New York, p 307
22. Chapiro A (1962) Radiation chemistry of polymeric systems. Wiley, New York
23. Ohnishi S, Ikeda Y, Kashiwagi M, Nitta I (1961) Polymer 2: 119
24. Waterman DC, Dole M (1970) J Phys Chem 74: 1906
25. Ichikawa T (1988) J Phys Chem 92: 1431
26. Ichikawa T, Yoshida H (1988) J Phys Chem 92: 5684
27. Iwasaki M, Toriyama K, Fukaya M, Muto H, Nunome K (1985) J Phys Chem 89: 5278
28. Ichikawa T, Kawahara S, Yoshida H (1985) Radiat Phys Chem 26: 731
29. Brede O, Neuman W (1988) Radiat Phys Chem 18: 1041
30. Libby W (1961) J Chem Phys 35: 1714
31. David C, Fuld D, Geuskens G (1979) Makromol Chem 132: 269
32. Schneider EE (1955) Disc Faraday Soc 19: 158
33. Abragam RJ, Melville HW, Ovenall DW, Wiffen DH (1958) Trans Faraday Soc 54: 1133
34. Ingram DJE, Symons MCR, Townsend MG (1958) Trans Faraday Soc 54: 409
35. Iwasaki M, Sakai Y (1969) J Polym Sci A1: 1079
36. Chen CR, Knight RL, Pollock L (1987) J Polym Sci Polym Chem Ed 25: 129
37. Ichikawa T, Yoshida H (1990) J Polym Sci Polym Chem Ed 28: 1185
38. Tabata M, Nilsson G, Lund A (1983) J Polym Sci Polym Chem Ed 21: 3257

39. Geuskens G, David C (1973) Makromol Chem 165: 273
40. Symons MCR (1963) Advan Phys Org Chem 1: 283
41. Torikai A, Kato R (1978) J Polym Sci Polym Chem Ed 16: 1487
42. Butyagin P Yu, Dubinskaya AM, Radtsig VA (1969) Russ Chem Rev 38: 290
43. Tanaka M, Yoshida H, Ichikawa T (1990) Polym J 22: 835
44. Huddson RL, Williams F (1978) J Phys Chem 82: 967
45. Plaček J, Szöcs F (1989) Eur Polym J 25: 1149
46. Ogasawara M, Tanaka M, Yoshida H (1987) J Phys Chem 91: 938
47. Muto H, Toriyama K, Nunome K, Iwasaki M (1984) Chem Phys Lett 105: 592
48. Burlandt W, Green D, Taylor C (1959) J Appl Polym Sci 1: 296
49. Todd A (1960) J Polym Sci 42: 223
50. Lehockey EM, Reid I, Hill I (1988) J Vac Sci Technol A 6: 2221

Editor: S. Okamura
Received December 11, 1991

Application of Pulse Radiolysis to the Study of Polymers and Polymerizations

M. Ogasawara

Hokkaido University of Education, Hakodate 040, Japan

Pulse radiolysis studies concerning the polymerization as well as the degradation, crosslinking and radiation resistance of polymers are surveyed. Initiation mechanisms of the radiation-induced polymerization of styrene and other monomers are discussed on the basis of the direct measurements of the reaction intermediates. Optical and kinetic data on the short-lived chemical intermediates produced in the solution of polymers and in the rigid polymers are surveyed and discussed with special reference to the degradation mechanism of polymers.

1 Introduction

Pulse radiolysis is a powerful tool for the creation and kinetic investigation of highly reactive species. It was introduced to the field of radiation chemistry at the end of the 1950s and became popular in the early 1960s. Although the objects of this modern technique were, at first, limited to solvated electrons and related intermediates, it was soon applied to a variety of organic and inorganic substances. As early as 1964, ionic intermediates produced by electron pulses in vinyl monomers were reported for the first time. Since then, the pulse radiolysis method has achieved considerable success in the field of polymer science.

The pulse radiolysis method is suitable for investigating dynamic processes induced by ionizing radiation in a system. Following the transfer of energy from radiation to medium, ionization and electronic excitation of constituent molecules occur. Molecules and ions excited to the levels above their dissociation limit undergo self-decomposition. Electrons ejected into the medium are thermalized by transferring their excess energy to the environment. Then the reactions between newly formed intermediates and their nearest neighbors occur in "spurs": the spur is a kind of small zone of excited and ionized species in the medium created by a packet of energy. Eventually, the spurs dissipate; any intermediates which escaped from the spur reactions are available for the reactions in the bulk solution. These processes take place in sequence with their time scale spanning 10^{-15} s to 10^2 s or even to days and years. Although the very early and late events are beyond the scope of this method, the "time window" of pulse radiolysis is wide enough to cover the chemically interesting processes such as in spur reactions and homogeneous reactions in the bulk.

Application of pulse radiolysis to polymers and polymerization was motivated at first by the success of radiation-induced polymerization as a novel technique for polymer synthesis. It turned out that a variety of monomers could be polymerized by means of radiolysis, but only a little was known about the reaction mechanisms. Early studies were, therefore, devoted to searching for initiators of radiation-induced polymerization such as radicals, anions and cations derived from monomers or solvents. Transient absorption spectra of those reactive intermediates were assigned with the aid of matrix isolation technique. Thus the initiation mechanisms were successfully elucidated by this method. Propagating species also were searched for enthusiastically in some polymerization systems, but the results were rather negative, because of the low steady state concentration of the species of interest.

Another motivation has relevance to the application of radiation chemistry to industry. High energy irradiation on poly(methyl methacrylate) and its analogues, for instance, results in the extensive degradation of the polymers. These polymers are often used under radiation in industry. Fundamental understanding of the degradation mechanism may be achieved by detecting and monitoring the short-lived reaction intermediates produced in the irradiated polymers. These species elude observation by the usual steady state methods.

Reactions of polymers in solution can also be studied by pulse radiolysis. In the solutions containing polymers, reactive intermediates produced by radiolysis may react with polymers. If the intermediates or the products are optically active, the direct observation of the reactions is feasible. Energy transfer and charge transfer processes in polymers are also interesting objects of pulse radiolysis.

From the author's view, the motivations for applying pulse radiolysis to polymer chemistry can be categorized into two groups. One is to understand the events induced in the polymers or polymerization systems which are exposed to ionizing radiation. By observing the reactive intermediates, one may clarify the effect of radiation on the polymers and polymerization from the mechanistic aspects. The available information is useful for exploring new techniques for polymer synthesis or creating radiation-sensitive or -resistant polymers. Second is to elucidate the dynamic aspects of polymers, in the sense of reactivity towards intermediates. In some cases, the effect of radiation on the systems is not really a main concern, but the radiation is used just for producing reactive intermediate which are necessary for probing the diffusional processes or conformational motion of polymers.

It is from these perspectives that we have reviewed the pulse radiolysis experiments on polymers and polymerization in this article. The examples chosen for discussion have wide spread interest not only in polymer science but also in chemistry in general. This review is presented in six sections. Section 2 interprets the experimental techniques as well as the principle of pulse radiolysis; the description is confined to the systems using optical detection methods. However, the purpose of this section is not to survey detail techniques of pulse radiolysis but to outline them concisely. In Sect. 3, the pulse radiolysis studies of radiation-induced polymerizations are discussed with special reference to the initiation mechanisms. Section 4 deals with applications of pulse radiolysis to the polymer reactions in solution including the systems related to biology. In Sect. 5 reaction intermediates produced in irradiated solid and molten polymers are discussed. Most studies are aimed at elucidating the mechanism of radiation-induced degradation, but, in some cases, polymers are used just as a "medium" for short-lived species of chemical interest. We conclude, in Sect. 6, by summarizing the contribution of pulse radiolysis experiments to the field of polymer science.

2 Principles and Technique of Pulse Radiolysis

2.1 Principles

Pulse radiolysis is one of the experimental methods for studying chemical kinetics. It is characterized by a short-lived perturbation of a system from the

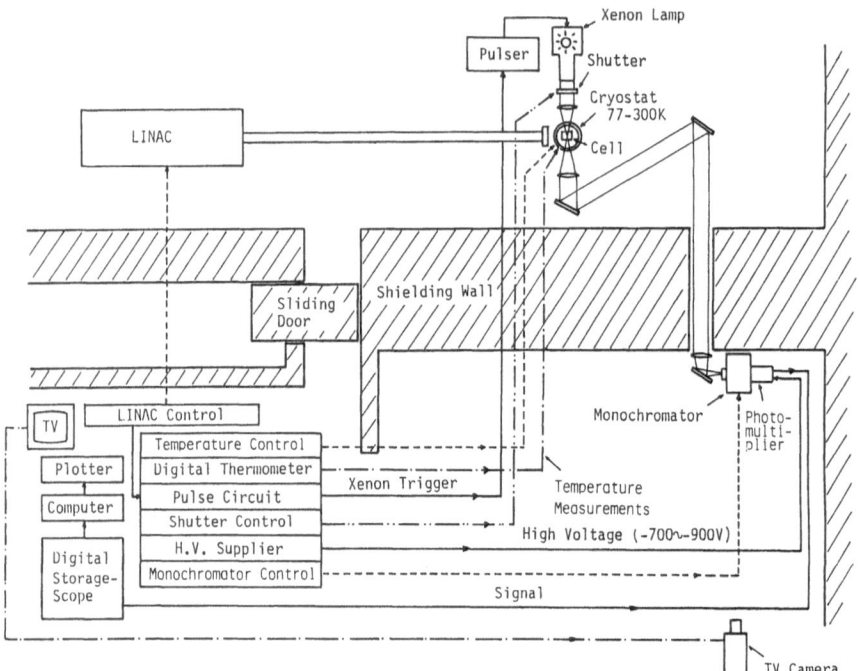

Fig. 1. Block diagram of the nanosecond pulse radiolysis system using the Hokkaido University 45 MeV electron linear accelerator

external radiation source followed by the direct observation of the relaxation of the system. The technique used is essentially the same as that of flash photolysis or laser photolysis except for the external energy source. Perhaps it may be considered a radiation chemical version of flash photolysis.

The basic arrangement of the pulse radiolysis system is shown schematically in Fig. 1. The sample in an absorption cell is irradiated by short pulses of ionizing radiation. The transmission of the sample can be measured by means of a light beam as a function of time; chemical reactions induced by the ionizing radiation are detected by the change in transmission. The emission of the sample can easily be measured by the similar setup except for the analyzing light source being turned off. Besides the optical measurement, electron spin resonance absorption, electrical conductivity, and light scattering intensity measurement can be used for detecting the transients [1–3].

2.2 Radiation Sources

Ideally the pulse of radiation for studying the evolution and decay of short-lived intermediates through time-resolved spectroscopy should be a δ-function or a step function electron beam. It should be high enough in intensity to generate a strong signal but low enough to avoid unnecessary damage to the sample. In practice, an electron pulse has a certain duration time, to which the rise and decay times are usually correlated. To meet requirements of the study of ultra-fast physical and chemical phenomena, it is essential to reduce the duration time of the pulse to a level which is negligible in comparison with the time-scale of the reaction under investigation. Considerable efforts have been made along this line on improving machines for electron acceleration.

So far the microwave electron linear accelerator is the most suitable for this purpose. In this accelerator electrons are injected into an evacuated cylindrical waveguide in which pulsed radiofrequency of several megawatts from a klystron oscillator travels. Electrons enter the radiofrequency field at the correct phase are accelerated to a velocity close to that of light. By means of gun control, electrons are injected only during the radiofrequency pulse, and thus the electron pulses of several nanosecond duration, useful for conventional nano-second or microsecond pulse radiolysis, are produced.

Other electron accelerators, such as Van de Graaff electron accelerators and the Febetron, are useful alternatives to the microwave linear accelerator. The Van de Graaff accelerator consists of a metal hemisphere suspended on the upper space of the accelerator in a mixed nitrogen-carbon dioxide gas environment. Electric charge is supplied by a rotating belt from a high voltage electric source placed at the bottom of the accelerator. The accelerated electron beam is produced in either continuous or pulsed by the discharge of the metal hemi-sphere toward the ground. They have an upper limit due to the space charge of about 4 ampere and are uneconomical above 3–4 MeV. The Febetron is a kind of an impulse generator consists of a number of capacitors. A high voltage pulse is applied to an accelerating tube to produce surge current of many thousands of ampere. Since operation at an energy greater than 1 MeV is difficult, a certain trick is necessary to get uniform irradiation even on a sample cell as thin as 2 mm.

2.3 Optical Detection Systems

Detection systems for nanosecond and microsecond pulse radiolysis are mostly the same, but that for picosecond measurement is somewhat different. The description is confined here to nanosecond and microsecond systems and a brief explanation to the picosecond system will be given in the following subsection.

As shown in Fig. 1, the light from a Xenon lamp is passed through the irradiation cell once, or many times if necessary, by a suitable mirror arrange-ment and focused on the entrance slit of the monochromator. The mono-chromator is separated from the radiation source and protected from the radiation. The light after the monochromator is finally monitored by a photo-multiplier at a selected wavelength. The electric signal from the photomultiplier is brought into a storage oscilloscope to be digitized and displayed. The data are processed, if necessary, by a computer linked to the oscilloscope and usually stored on floppy disks.

To attain a resolved time of a few nanoseconds with a conventional photomultiplier, the anode is coupled directly to a oscilloscope through 50-ohm coaxial cable; special attention should be paid to attaining impedance matching. Photomultiplier tubes should be selected carefully. Their response to a short pulse of light is not necessarily clean; occasionally the tail of the impulse signal lasts longer depending on the tube [4]. Generally speaking, the photomultiplier of head-on type has a longer response time in comparison with that of the side-on type.

One of the most important factors determining the signal to noise ratio S/N is the so called shot noise which arises from statistical fluctuations in the photon flux at the cathode. In the case where the shot noise is predominant, the S/N is proportional to the root of the number of photoelectrons per second, provided other parameters such as rise time of the photomultiplier are kept constant. Therefore, for obtaining better S/N ratio, the input signals should be as strong as possible within the range in which the photomultiplier output is linear to the input signals. Electric circuits for the photomultiplier are designed to obtain a wider linear output range: each dynode of the photomultiplier is linked by a graded series of storage capacitors to keep the interdynode voltage constant for high anode current for a certain time (usually several millisecond). The voltage applied to each dynode should be high to reduce the space charge limitation.

2.4 Picosecond Pulse Radiolysis Systems

Pulse radiolysis systems capable of picosecond time resolution use the fine structure of the output from the electron linear accelerator. Electrons in the accelerating tube respond to positive or negative electric field of the radiofrequency, and they are eventually bunched at the correct phase of the radiofrequency. Thus the electron pulse contains a train of bunches or fine structures with their repetition rate being dependent on the frequency of the radiofrequency (350 ps for the S-band and 770 ps for the L-band).

In stroboscopic picosecond pulse radiolysis the train of bunches is used as it is obtained from the electron linear accelerator [5]. A set of Čerenkov light flashes is produced when the fine structure electron pulses travels through the air before hitting a sample; the flash is used to measure the concentration of the absorbing species produced in the sample at a finite point in time. The light has a continuous spectrum, increasing in intensity toward the ultraviolet, and it has the same time-profile as the fine structure electron pulse. It should be noted that in this detection system the optical detector does not (and needs not) resolve individual analyzing-light flashes but integrates the total number of photons. The time sweep, which is limited within the interval between fine structure pulses though, is accomplished by varying the length of the optical delay path.

With the aid of a subharmonic prebunching cavity operated at a few tenths of the main accelerating microwave frequency, the electron linear accelerator is

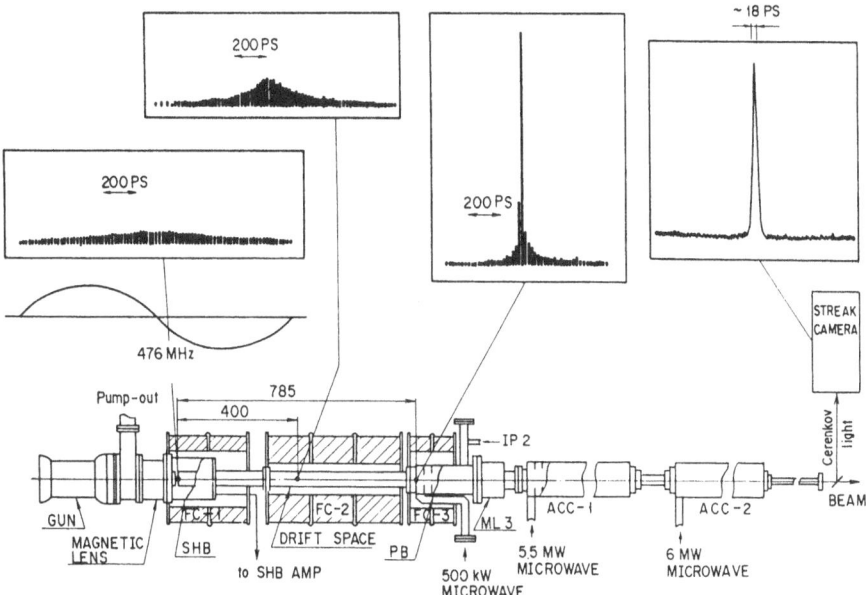

Fig. 2. Bunching process of electron beam by the subharmonic prebuncher operated at 476 MHz [7]

able to produce a picosecond single beam. The principle is that the velocity of the electron beam emitted from an electron gun is modulated by the electric field at the gap of the prebuncher to form a single bunch within one radiofrequency period as illustrated in Fig. 2 [6, 7]. The optical detection setup for single bunch picosecond pulse radiolysis is not much different from that for nanosecond pulse radiolysis except for the streak camera which is used as a photo-detector [8]. In the picosecond measurement the number of photons from the analyzing light source is so small, despite the intense light flashes from the pulsed Xenon lamp, that the disturbing Čerenkov light from the irradiated sample cannot be neglected. The contribution of the Čerenkov light, measured in the absence of the analyzing light, is eliminated from the time-profile of the light intensity by using a computer [8].

3 Vinyl Monomers

3.1 Polymerization of Styrene and α-Methylstyrene

3.1.1 Early Studies

When vinyl monomers such as isobutene, styrene, α-methylstyrene, and vinyl ethers are irradiated by ionizing radiation, the polymerizations of these monomers take place [9]. The radiation-induced primary events in these systems drew much attention in conjunction with the initiation mechanisms of the polymerizations. In 1965, one of the first pulse radiolysis studies of the polymerization was reported by Katayama et al.; the experiment was aimed at detecting initiators of the radiation-induced polymerization of α-methylstyrene [10]. The observed transient absorption in the UV region was assigned to the radical anion of α-methylstyrene produced by the electron capturing reaction of the monomer [10–12]. Similar results have been obtained by Schneider and Swallow with both styrene and α-methylstyrene [13]. A following comprehensive study by Metz et al. [14] disclosed the details of the transients produced in the pulsed neat styrene. The absorption, having a maximum at 380 nm and a lifetime of a few microseconds, was assigned to the radical anion of styrene. The steady-state absorption spectrum of styrene radical-anion, observed with γ- irradiated 2-methyltetrahydrofuran (MTHF) glasses containing styrene at 77 K, had a strong band at 410 nm ($\varepsilon \sim 23\,000\,\text{mol}^{-1}\,\text{dm}^3\,\text{cm}^{-1}$) and a relatively weak one at 600 nm ($\varepsilon \sim 5000\,\text{mol}^{-1}\,\text{dm}^3\,\text{cm}^{-1}$) [15]. Thus the formation of the monomeric radical anions was confirmed in the irradiated styrene.

The transient absorption of the radical anions observed in pulsed styrene and α-methylstyrene were extremely sensitive to water: they were greatly diminished or sometimes not observed at all if a small amount of water, even moisture in the atmosphere, was introduced into the sample. Similar phenomena have been observed in the pulse-irradiated monomers in cyclohexane solutions [16, 17]. Addition of ethanol, methylene chloride, chloroform, carbon tetrachloride, and n-butyl amine also reduced the yield of the anions [18].

On the other hand, it was known that the radiation-induced polymerizations took place very efficiently, only when the monomers were dried rigorously [19]. The polymerization rate was found to be 100 times higher than that in "wet" systems where the radical mechanism was predominant [11, 14]. On the basis of the parallel effects of water to the polymerization rate and the short-lived intermediates, Katayama et al. concluded that the radical anion might play a role in the initiation stages of the radiation-induced polymerization [11, 12]. However, both the effects of additives on the polymerization rate and the composition of the copolymer strongly suggested a cationic mechanism [19, 20]. The initiating species of the radiation induced polymerization had been a matter of dispute for a long time.

High energy radiation generates ionic and free radical intermediates in styrene, and both ionic and radical polymerizations take place simultaneously. Therefore, the observation of neutral radicals related to the radical polymerization had been expected. The absorption of radicals with a maximum at 320–330 nm was actually observed in the pulsed styrene and α-methylstyrene; it decayed over a time interval of several hundred microseconds without being affected by the presence of water [12–14]. This absorption was attributed for the main part to a radical with a benzyl-type structure. The similar absorptions were observed in the cyclohexane solutions [21]. Swallow suggested that this intermediate would be formed by an ionic process, probably by the protonation of a styrene anion by its germinate partner [21]. The spectrum was overlapped by several other radicals including polymer radicals. Steady state measurements revealed that even a permanent absorption, composed at least of two overlapping peaks with apparent maxima at 320 nm and at 310 nm, was produced by radiolysis, although neither of them have yet been identified.

3.1.2 Formation and Reaction of Radical Cations

Although the polymerization experiments suggested the cationic mechanism, the behavior of cationic species in the irradiated styrene and α-methylstyrene had remained obscure for a long time. Yoshida et al. succeeded for the first time in detecting cations in pulse-irradiated n-butyl chloride containing styrene in 1971 [22]. An absorption band of radical cation ($\lambda_{max} = 350$ nm) produced by the positive charge transfer from the solvent to the solute styrene was observed in the UV region in addition to a broad absorption at about 460 nm; the latter band was attributed to the dimer cation. From the analysis of decay curves, the second-order rate constant for the reaction of styrene radical-cation with neutral styrene was estimated to be $\sim 2 \times 10^6$ mol^{-1} dm^3 s^{-1}. The rate constant of the propagation reaction, which refers the addition of styrene monomer to carbonium ion of the chain ends, is about 10^7 mol^{-1} dm^3 s^{-1} for the radiation-induced cationic polymerization of styrene. The difference may be caused by resonance effect of radical cations [22].

As for the positive charge transfer in n-butyl chloride, Mehnert et al. have proposed the following elementary processes [23]:

$$\text{BuCl} \xrightarrow{\hspace{0.5cm}} \text{BuCl}^{\dagger}, \text{Bu}^+, \text{HCl}, \text{Cl}^-, \text{Bu}^{\cdot}, \text{etc.} \tag{1}$$

$$\text{BuCl}^{*\dagger} \longrightarrow \text{BuCl}^{\dagger}(\xrightarrow{+\text{Cl}^-} \text{BuCl}^{\dagger} \cdots {}^-\text{Cl})$$

$$\longrightarrow \text{Bu}^+ + \text{HCl}$$

$$\xrightarrow{+\text{St}} \text{BuCl} + \text{St}^{\dagger}$$

$$\text{BuCl}^{\dagger} + \text{St} \rightarrow \text{St}^{\dagger} + \text{BuCl}$$

$$\text{Bu}^+ + \text{St} \rightarrow \text{St}^{\dagger} + \text{Bu}^{\cdot}$$

where St denotes the styrene monomer. Probably, vibronically-excited solvent-cations (BuCl*†) are formed as a primary product of the ionizing act, which either decompose or transfer their energy and charge to neighboring molecules.

Egusa et al. disclosed the behaviors of cationic intermediates in the initiation stage of the polymerization of styrene and α-methylstyrne by low temperature pulse radiolysis [24, 25]. Figure 3 shows the transient absorption spectra for styrene in a mixture of isopentane and n-butyl chloride (4:1 by volume) irradiated with 4 μs pulses at -165 °C. M_1 and M_2 refer the absorption bands

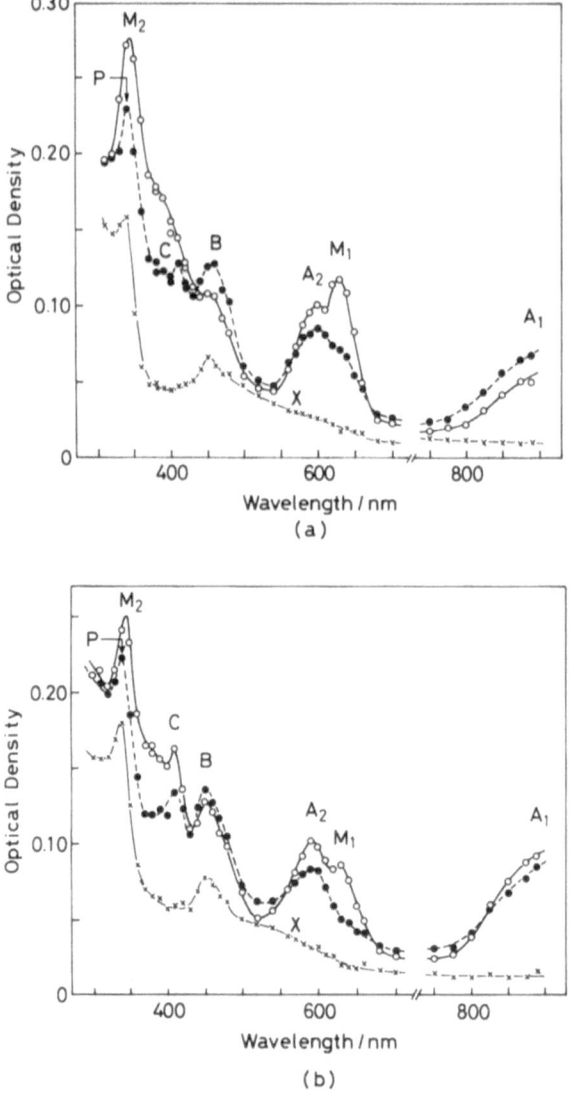

Fig. 3a, b. Transient absorption spectra for (**a**) 0.05 and (**b**) 0.15 mol dm^{-3} styrene solutions in isopentane-*n*-butyl chloride mixtures irradiated with 4 μs pulses at − 165 °C: (○) at the end, (●) after 6 μs, and (×) after 60 μs of an election-pulse [25]

due to monomeric radical cations. The observation of the band at 630 nm (M_1) had been expected from the results of steady-state absorption measurements of the styrene radical-cation in a rigid matrix [26]. A_1 in the near IR region and A_2 at 600 nm were found to have originated from the same species. Egusa et al. ascribed these two bands to the dimer cation of styrene (St_2^+) because of the characteristic charge-resonance absorption band in the near-IR region [24, 27].

Table 1. Assignment of absorption bands observed in pulse radiolysis of 1,1-diphenylethylene in dichloromethane

Absorption band (nm)	Assignment	Assumed formula of species
335	free radical	e.g. $\begin{array}{c} \text{ph} \\ | \\ \cdot\text{C}-\text{CH}_2-\text{R} \\ | \\ \text{ph} \end{array}$
395	dimeric radical cation (strong); monomeric radical cation (weak);	$\left(\begin{array}{c} \text{ph} \\ | \\ \text{C}=\text{CH}_2 \\ | \\ \text{ph} \end{array}\right)^{\ddagger}_2, \quad \left(\begin{array}{c} \text{ph} \\ | \\ \text{C}=\text{CH}_2 \\ | \\ \text{ph} \end{array}\right)^{\ddagger}$
435	growing cation (bonded dimer)	$\begin{array}{c} \text{ph} \\ | \\ {}^{+}\text{C}-\text{CH}_2-\text{R} \\ | \\ \text{ph} \end{array}$
550	monomeric radical cation	$\left(\begin{array}{c} \text{ph} \\ | \\ \text{C}=\text{CH}_2 \\ | \\ \text{ph} \end{array}\right)^{\ddagger}$
1000–1200	dimeric radical cation	$\left(\begin{array}{c} \text{ph} \\ | \\ \text{C}=\text{CH}_2 \\ | \\ \text{ph} \end{array}\right)^{\ddagger}_2$

The dimer cation was supposed to have a sandwich structure in which the orbitals of one molecule overlapped with those of the other molecule. The band at 450 nm (B) is due to the bonded dimer cation (St–St‡); the formation of this species corresponds to the initiation step of the polymerization. The bonded dimer cation may be formed by the opening of the vinyl double-bonds. Egusa et al. proposed that the structure was a linked head-to-head type I or II, by the analogy of the dimeric dianions of styrene and α-methylstyrene. Table 1 summarizes the assignment of absorption bands observed in pulse radiolysis of 1,1-diphenylethylene in dichloromethane, which is a compound suitable for studying monomeric and dimeric cations [28].

$$I: \left[\begin{array}{c} \text{CH—CH}_2\text{—CH}_2\text{—CH} \\ \bigcirc \qquad\qquad \bigcirc \end{array} \right]^{\dot{+}} \qquad II: \left[\begin{array}{c} \text{CH}_3 \qquad\qquad \text{CH}_3 \\ \text{C—CH}_2\text{—CH}_2\text{—C} \\ \bigcirc \qquad\qquad \bigcirc \end{array} \right]^{\dot{+}} \tag{2}$$

Most of the dimer cations and the bonded dimer cations were produced during the pulse irradiation. This can be explained by assuming the existence of the equilibrium between styrene monomers (St) and neutral dimers of styrene (St$_2$) under the experimental conditions [24].

$$St + St \rightleftharpoons St_2 \tag{3}$$

Both the styrene monomer and the neutral dimer can trap a migrating positive hole or positive charge from solvent radical-cations (solvent$^{\dot{+}}$) or related cationic species, which leads to the formation of radical cations, dimer cations, and bonded dimer cations.

$$\text{solvent}^{\dot{+}} + St_2 \rightarrow \text{solvent} + St_2^{\dot{+}}$$
$$\text{solvent}^{\dot{+}} + St_2 \rightarrow \text{solvent} + \text{St–St}^{\dot{+}} \tag{4}$$

Part of the dimer cation and the bonded dimer cation were formed relatively slowly by the following reactions.

$$St^{\dot{+}} + St \rightarrow St_2^{\dot{+}}$$
$$St^{\dot{+}} + St \rightarrow \text{St–St}^{\dot{+}} \tag{5}$$

The propagation stages of the cationic polymerization are not well understood. Egusa et al. have claimed that the long-lived absorption at 340 nm observed from the styrene solution in n-butyl chloride might be due to the bonded-trimer radical cation $\cdot$St–St–St$^+$ which was formed by the reaction of a bonded dimer cation with a styrene monomer [24].

Mehnert et al. studied the formation of cationic intermediates in alkane solutions containing a small amount of electron scavengers such as carbon tetrachloride [29, 30]. The radical cations of styrene are formed by the following mechanism:

$$S \rightsquigarrow S^{\dot{+}} + e^-$$
$$e^- + CCl_4 \rightarrow \cdot CCl_3 + Cl^-$$
$$S^{\dot{+}} + St \rightarrow St^{\dot{+}} + S$$
$$St^{\dot{+}} + Cl^- \rightarrow \text{product} \tag{6}$$

The reaction of the primary solvent cations (or holes) with monomers to yield styrene radical cation is very fast ($k \sim 10^{11}\,\text{mol}^{-1}\,\text{dm}^3\,\text{s}^{-1}$). The produced radical cations are consumed by the following reactions:

$$St^{+} + St \rightarrow St_{2}^{+}$$
$$St_{2}^{+} + Cl^{-} \rightarrow product$$
$$St_{2}^{+} + St \rightarrow St_{3}^{+}$$
$$St_{2}^{+} + A \rightarrow product \tag{7}$$

For the first several hundred nanoseconds after the pulse, the decay curves of the radical cations and the dimer radical-cation deviated from a usual first- or second-order kinetics. Mehnert et al. analyzed the curves in the region of inhomogeneous kinetics on the basis of the pair recombination concept of Rzas et al. [31].

The behavior of cationic intermediates produced in styrene and α-methyl-styrene in bulk remained a mystery for a long time. The problem was settled by Silverman et al. in 1983 by pulse radiolysis in the nanosecond time-domain [32]. On pulse radiolysis of deaerated bulk styrene, a weak, short-lived absorption due to the bonded dimer cation was observed at 450 nm, in addition to the intense radical band at 310 nm and very short-lived anion band at 400 nm (Fig. 4). (The lifetime of the anion was a few nanoseconds. The shorter lifetime of the radical anion compared with that observed previously may be due to the different purification procedures adopted in this experiment, where no special precautions were taken to remove water). The bonded dimer cation reacted with a neutral monomer with a rate constant of $10^{6} \, mol^{-1} \, dm^{3} \, s^{-1}$. This is in reasonable agreement with the propagation rate constant of radiation-induced cationic polymerization.

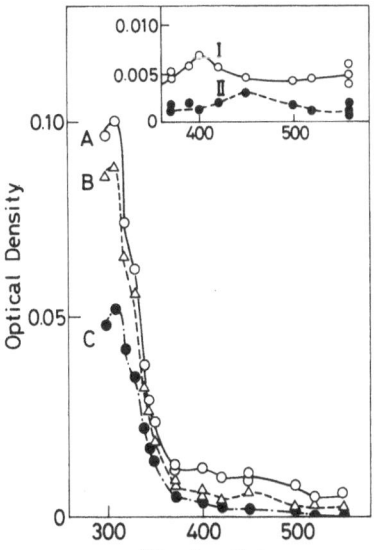

Fig. 4. Absorption spectra from deaerated bulk styrene observed (A) immediately, (B) at 10 ns, and (C) at 100 ns after a 5 ns electron-pulse. Insert: spectra I (A-B) very short-lived intermediates and II (B-C) short-lived intermediates [32]

It was found in this experiment that both anionic and cationic species reacted efficiently with methanol in bulk styrene. The bonded dimer cations and the radical anions were converted to long-lived benzyl radicals, which initiated the radical polymerization. The G value of the propagating benzyl radical was only 0.7 in pure styrene, but it increased up to 5.2 in the presence of methanol. A small amount of methanol converted almost all the charge carriers to propagating free radicals; this explains why the mechanism of radiation-induced polymerization is changed drastically from cationic to radical processes on adding methanol.

3.1.3 Effects of Counterions in Ionic Polymerizations

Apart from the relevance to the radiation-induced polymerizations, the pulse radiolysis of the solutions of styrene and α-methylstyrene in MTHF or tetrahydrofuran (THF) has provided useful information about anionic polymerization in general [33]. Anionic polymerizations initiated by alkali-metal reduction or electron transfer reactions involve the initial formation of radical anions followed by their dimerization, giving rise to two centers for chain growth by monomer addition [34]. In the pulse radiolysis of styrene or α-methylstyrene (MS), however, the rapid recombination reaction of the anion with a counterion necessarily formed during the radiolysis makes it difficult to observe the dimerization process directly. Langan et al. used the solutions containing either sodium or lithium tetrahydridoaluminiumate (NAH or LAH) in which the anions formed stable ion-pairs with the alkali-metal cations whereby the radical anions produced by pulse radiolysis could be prevented from rapid recombination reaction [33].

$$NaAlH_4 \rightleftharpoons Na^+ + AlH_4^-$$

$$Na^+ + MS^{\overline{\cdot}} \rightarrow (Na^+, MS^{\overline{\cdot}}) \tag{8}$$

The dimerization of the anion radicals was studied in NAH/THF and LAH/THF by following the decay of the absorption maxima of the styrene, α-methylstyrene, and 1,1-diphenylethylene radical-anions. The dimerization followed second order kinetics with the rate constants shown in Table 2. The absorption spectrum with a peak at 320 nm, which grew concomitantly with the decay of the ion-pair $MS^{\overline{\cdot}}$, Na^+(or Li^+) was assigned to the dimeric dianion $^-MSMS^-$ in the form of the solvent-separated ion pair (Fig. 5).

$$2(Na^+, MS^{\overline{\cdot}}) \rightarrow Na^+, {}^-MSMS^-, Na^+ \tag{9}$$

Another investigation along this line is the pulse radiolysis study of the electron transfer reactions from aromatic radical anions to styrene; this type of reaction is commonly used to initiate anionic polymerization of styrene [35]. The electron transfer rates from the unassociated biphenyl radical-anions to styrene derivatives in 2-propanol were found to increase along the

Table 2. Rate constants for the dimerization of vinylic ion pairs

Ion pair	Rate constant/ 10^8 mol^{-1} dm^{3}s^{-1}	Dose range/Gy
(Li$^+$, D$^{\cdot}$)	15.5 ± 1.2	45–180
(Na$^+$, D$^{\cdot}$)	6.0 ± 0.3	45–180
(Na$^+$, D$^{\cdot}$)	5.0	—
(Li$^+$, MS$^{\cdot}$)	12.3 ± 0.3	80–240
(Na$^+$, MS$^{\cdot}$)	6.6 ± 0.1	80–240
(Na$^+$C, MS$^{\cdot}$)	64.2 ± 0.6	80, 135
(Li$^+$, St$^{\cdot}$)	10.9 ± 0.5	60–180
(Na$^+$, St$^{\cdot}$)	10.0 ± 0.2	60–180

D: 1,1-diphenylethylene; MS: α-methylstyrene; St: styrene; C: crown ether

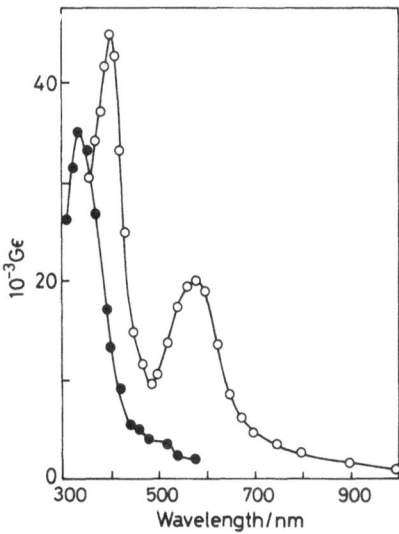

Fig. 5. Transient absorption spectra observed in saturated solution of tetrahydrofuran with sodium tetrahydridaluminate containing 5×10^{-2} mol dm^{-3} α-methylstyrene: ○, (Na$^+$, MS$^{\cdot}$); ●, Na$^+$, $^-$MSMS$^-$, Na$^+$ [33]

series *p*-methoxystyrene < styrene < *p*-methylstyrene < *m*-chlorostyrene < *p*-chlorostyrene as shown in Fig. 6; this is compatible to the prediction of Marcus theory. But the rate of the electron transfer from the ion-pair consists of the biphenyl radical-anion and Na$^+$ to α-methylstyrene, styrene, *p*-methoxy-styrene, and *p*-chlorostyrene did not follow the change of the reduction potentials of the electron acceptors. This abnormality is obviously due to the effect of the Na$^+$ associated with the anion.

Mah et al. demonstrated the effect of counterions on the cationic polymerization of styrene [35–37]. The radiation-induced polymerization is much more sensitive to impurities than the catalytic polymerization, as the former involves the cationic species in a free ion state. Thus, one can expect, in the presence of stable anions, the promotion of the cationic polymerization because of the ion-pair formation between the dimer cation and the counterion. The effect was

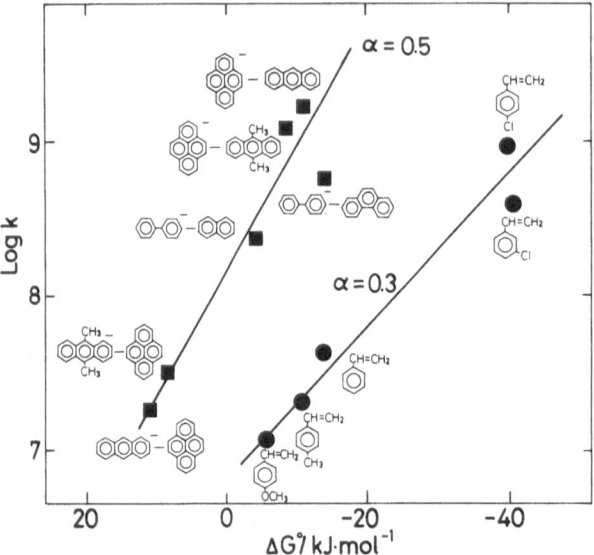

Fig. 6. Correlation of the electron transfer rate constant with free energy change of the electron transfer reactions. Both the donors and the acceptors are illustrated for the arene⁻-arene series, but only the acceptors are illustrated for the biphenyl-styrene series. α is the gradient of the straight line [35]

actually observed in the radiation-induced polymerization of styrene; the addition of triphenyl-sulfonium hexafluorophosphate $((C_6H_5)_3SPF_6)$ promoted the polymerization in methylene chloride. From the results of the pulse radiolysis measurements, Mah et al. concluded that the promotion should be related to the increase in both the yield and the lifetime of the styrene dimer-cation [36]. They attributed the increase of the lifetime to the stabilization of the dimer cation against Cl⁻. They also studied the effect of other additives, such as $(C_6H_5)_2ICl$ [37], $(C_6H_5)_2IPF_6$ [37] and $(n\text{-}C_4H_9)NPF_6$ [38] in a series of experiments.

3.2 Polymerization and Dimerization of N-Vinylcarbazole

Next to the styrene compounds, N-vinylcarbazol has been most extensively studied by pulse radiolysis. When N-vinylcarbazole is irradiated in aerated benzonitril, the cyclodimer of the N-vinylcarbazole is formed; the polymer is formed in both aerated and deaerated nitrobenzene. Tagawa et al. have proved that the radical cation of N-vinylcarbazole plays an important role in both cyclodimerization and polymerization processes [39–43].

The spectrum of the monomeric radical cation of N-vinylcarbazole had a sharp peak at 790 nm and broad ones at 700, 640, and 510 nm [42]. The G value of the radical cation was only 0.8, but that for the formation of the dimer, *trans*-1,2-dicarbazylcyclobutane, reached the order of several hundreds. Thus the

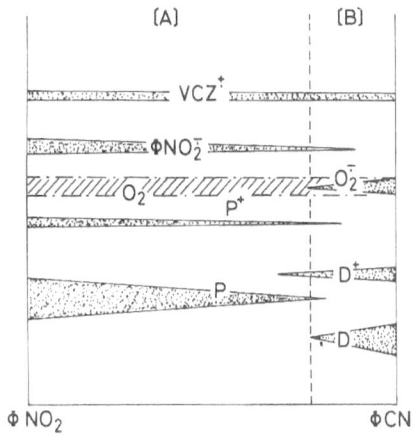

Fig. 7. Intermediate and product species formed in the dimerization and the polymerization of vinylcarbazole in nitrobenzene (ΦNO_2) and benzonitrile (ΦCN) in the presence of oxygen: ($VCZ^{\pm}$) monomer cation radical; (ΦNO_2^-) nitrobenzene anion; ($D^{\pm}$) dimer cation radical; (P^+) polymer cation; (D) cyclodimer; (P) polymer [43]

existence of a chain reaction was confirmed [39]:

$$S \rightsquigarrow S^{\pm}, e^-, \text{ other products}$$

$$S^{\pm} + M \rightarrow S + M^{\pm}$$

$$M^{\pm} + M \rightarrow M_2^{\pm}$$

$$M_2^{\pm} + M \rightarrow M^{\pm} + M_2$$

$$M^{\pm} + X^- \rightarrow M + X$$

$$M_2^{\pm} + X^- \rightarrow M_2 + X \tag{10}$$

where S, M, and M_2 refer to a solvent molecule, a monomer, and a dimer molecule, respectively. X may be O_2, because the cyclodimerization occurs only in the presence of air. An oxygen molecule is necessary for the formation of the cyclodimer not only as chain carrier but also for preventing addition polymerization by forming a complex with the monomer cation.

In the aerated solution of N-vinylcarbazole in benzonitril, the observed spectrum was similar to that of the dimer cation, whereas the spectrum observed in nitrobenzene was comprised of monomer cation and polymer cation [43]. This implies the formations of cyclodimer and polymer through the dimer cation and the polymer cation, respectively. From the measurements with the mixed solvents of benzonitril and nitrobenzene, the relative amounts of intermediate species and final products were obtained as functions of solvent composition. The results are illustrated by Fig. 7 [43].

3.3 Pulse Radiolysis of Other Monomers

Only a few monomers, other than styrene and related monomers, have been investigated by pulse radiolysis. The studies on methyl methacrylate and related monomers have shown the formation of the associated dimer anion of the monomers. At low temperatures, the intense IR absorption attributable to dimer anion was found in the pulsed MTHF solution [44]. The dimer anion was formed by the reaction between methyl methacrylate and its radical anion. The similar bands were observed for fumaronitrile, maleic anhydride, citraconic anhydride, dimethyl maleic anhydride, methyl vinyl ketone, acrolein, crotonitrile, methacrylonitrile, methyl acrylate, and methyl crotonate. The results indicated that the electron affinity of a functional group was an important factor for the dimer-anion formation.

Ogasawara et al. studied nitroethylene in MTHF and tried to detect the dimer anion of it [45]. However, the observed absorption spectrum, having a maximum at 440 nm, was not due to the dimer anion but the radical anion of nitroethylene. The absorption in the region from 500 to 650 nm increased gradually with the decay of the nitroethylene radical-anion after a pulse; this absorption was tentatively assigned to the bonded dimer-anion.

4 Polymers in Solution

4.1 Polymer Ions in Solution

Primary processes induced by ionizing radiation in the solution are excitation and ionization of the solvent molecules. Subsequent electron attachment to solute polymers leads to the formation of polymer anions. So far the radical anions of poly(methyl methacrylate) (PMMA) [46, 47], substituted PMMA [46, 47], poly(4-vinylbiphenyl) (PVB) [47–50], poly(1-vinylpyrene) (PVP) [50], organopolysilane [51] and substituted polyacetylene [52] have been studied.

Ogasawara et al. found that the radical anions of PMMA, poly(ethyl methacrylate), poly(n-butyl methacrylate), and poly(isobutyl methacrylate) were generated by the reaction of e_s^- with polymers in hexamethylphosphorictriamide [46]. The second-order rate constant for the reaction between e_s^- and PMMA was evaluated as $7.7 \times 10^8 \, mol^{-1} \, dm^3 \, s^{-1}$ assuming homogeneous distribution of the reaction sites. The absorption spectra of the radical anions had a strong absorption band at $< 300 \, nm$ and a shoulder at 450 nm. The ketyl-type anion is the most probable structure for the PMMA anion:

$$\sim CH_2 - C(CH_3) \sim$$
$$|$$
$$\cdot CO^-$$
$$|$$
$$OCH_3$$

This structure was confirmed by the steady state optical measurements: the similar spectrum was observed from the radical anions of methyl isobtyrate, a model compound of PMMA, in γ-irradiated MTHF matrices.

Tachikawa et al. made ab initio MO calculations on the model compound of PMMA. Figure 8 shows the contour maps of unpaired-electron orbital for the radical anion of methyl isobutyrate with a geometrically optimized structure; the contour map of the anion obtained with a fixed geometry of the neutral molecule is also shown for comparison. The C–O bond of the ether part is extended by 0.04 nm on adding the excess electron to the molecules [53]. The excess electron enters the antibonding orbital of the carbonyl group, which is coupled with the 2s-orbital of C atom of the carbonyl group to form an sp^3 hybrid-orbital. This explains the considerable deformation of the negatively-charged molecule. The results of spin population analysis showed that the unpaired electron was distributed mostly on the carbonyl group. The ab initio calculations also suggested that the strong absorption in UV should be ascribed to two different allowed π–π^* transitions; the weak absorption in the visible (400–500 nm) was assigned to the n–π^* transition which was partially allowed owing to the bent configuration of the radical anions.

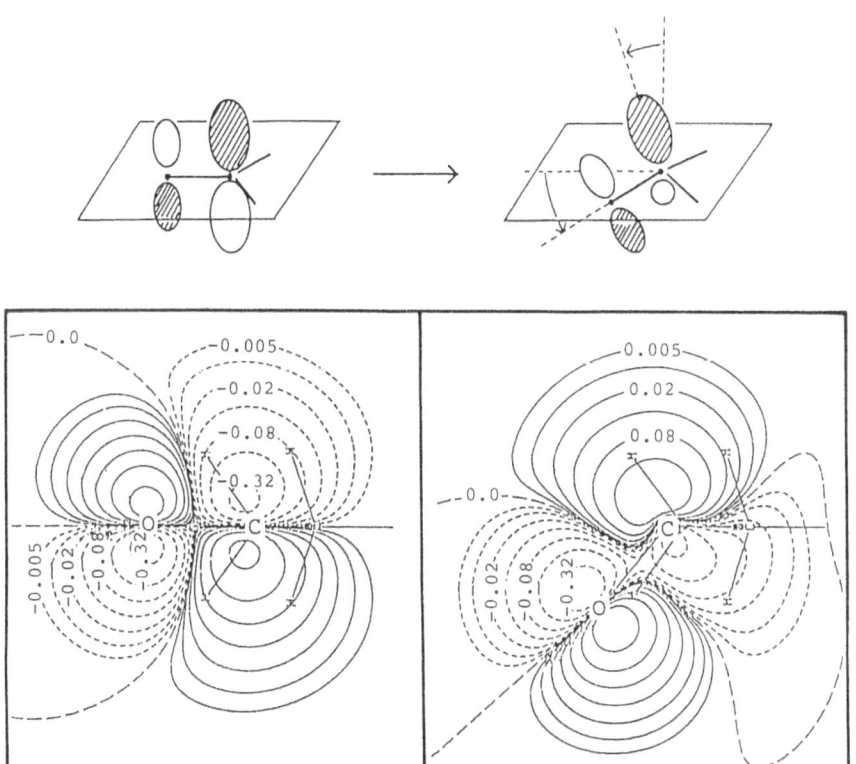

Fig. 8. Schematic representations and contour maps of the molecular orbitals on the carbonyl part of the methyl isobutyrate radical-anion with the fixed geometry of neutral molecule (*left*) and the fully optimized geometry (*right*).

The e_s^- reacts with PVB to give a polymer anion with a high efficiency [47]. The rate constant was evaluated as $4.7 \times 10^9 \, mol^{-1} \, dm^3 \, s^{-1}$ in hexamethylphosphorictriamide. The absorption specra of the radical anions of PVB [47] and PVP [48] are similar to those of biphenyl anion and pyrene anion, respectively, to mean that the excess electrons trapped by the polymers are essentially localized on the side groups.

Tanaka et al. studied the decay reactions of PVB radical anions produced by electron pulses in MTHF [47]. At low concentration (< 0.05 base-mol dm^{-3}) of polymers the decay reaction followed a simple second-order kinetics. The charge neutralization reaction is responsible for the decay curve as is the case of biphenyl radical anions. However, the rate constant of the polymer anions was only a half or one-third of that of the biphenyl anion, because of the small diffusion coefficient of the polymer ion in solution. At high concentration of the polymer, a spike was observed in the time-profile of the PVB anion; this was attributed to the retarded geminate recombinations within micro-domains where the polymers were entangled with each other.

The radical anions of organopolysilanes are unique in the distribution of the excess electron. Introducing a methylphenylsilane unit or diphenylsilylene unit into the polymer caused a red shift in the UV absorption of their radical anions [51]. This implies the existence of electronic interaction between the silicon main chain and the side group; namely, the excess electron is dispersed on the pendant phenyl group as well as the main chain. The delocalization of the excess charge in the organopolysilanes can be understood on the basis of the band model of Si-compounds. Ab initio calculations by Nelson and Pietro at the HF/3-21G level suggested the existence of $n - 1$ LUMO states in the polysilanes; where n refers the degree of polymerization [54]. The orbitals were composed primarily of in-phase combinations of out-of-plane p orbitals on each Si atom and only partially contributed by the anti-bonding interaction of Si–H. The excess electron was supposed to enter into these orbitals and distributed over the polymers.

The structure of the radical anions of substituted polyacetylene is not clear, although the pulse radiolysis study on these polymers in solution is now in progress [52]. In this relevance, the absorption spectra of the radical anions of retinal homologues are interesting [55]. The absorption maximum shifted to the lower energy side with increasing number of the double bond in the polyene from 1 to 9; obviously the negative charge was dispersed over the chain of the polymer.

The positive charge of solvent radical-cations transfers to solute molecules in halogenated hydrocarbons such as chloroform and dichloroethane. However, only few studies have been made on the radical cations of polymers in solution. Tanaka et al. observed the dimer cation of the biphenyl group or the pyrenyl group of the polymers in the pulse radiolysis of PVB and PVP in 1,2-dichloroethane [49]. The absence of the monomeric cation is due to the rapid intramolecular dimerization of the radical cations of the side groups of the polymers. Irie et al. observed two kinds of intramolecular dimer cations in the solution of 1,3-di(2-napthyl)propane and poly(2-vinylnaphthalene) and only one dimer cation in the solution of 1,12-di(2-napthyl)dodecane [56]. Conversion of one form of the dimer cation to the another was explained in terms of geometrical structure of the dimer cations.

In cyclohexane geminate recombination occurs very efficiently and the observation of polymer ions is rather difficult [57, 58]. However, when the electron scavenger such as chloroform and carbon tetrachloride was added to the solution of polystyrene in cyclohexane, a weak, broad absorption band with a maximum at 1000 nm due to dimer cation of benzene was observed. The dimer cation radical might be produced by the hole migration, along the polymer chain, from a radical cation to a site suitable for the dimer-cation formation [59].

4.2 Intramolecular Charge Transfer in Polymer Ions

The electron transfer from aromatic radical anions to various electron acceptors takes place efficiently in solution. Likewise, when a second solute, pyrene, is added to the MTHF solution of PVB, the electrons transfer from polymer anions to pyrene occurs [50]. The rate constant determined by pulse radiolysis is approximately a third of that of the electron transfer from biphenyl anion to pyrene.

The PVB anion, produced by electron pulses in MTHF, simply disappears by the neutralization reaction with the solvent cation. However, it has been shown that the excess electron trapped in PVB transfers to the neighboring side groups of the polymers before the disappearance of the polymer anion. Tanaka et al. synthesized poly(4-vinylbiphenyl-co-1-vinylpyrene) in which a small number of biphenyl side groups were substituted by pyrenyl groups; they expected a drastic spectral change due to the loss of biphenyl anion and the formation of pyrenyl anion when the excess electron residing in one of the biphenyl groups hopped to a pyrenyl group in the polymer [60]. Pyrenyl groups were chosen as indicator, because the spectral overlap between anion radicals of biphenyl and pyrene was the minimum among possible combinations of aromatic compounds.

Figure 9 shows the transient absorption spectra observed on pulse radiolysis of MTHF solution containing the copolymers. In this copolymer 0.56% of the biphenyl groups were substituted by pyrenyl groups. The changes in the concentration of the radical anions of biphenyl groups (Bp^{-}) and pyrenyl groups (Py^{-}) are shown by the changes in the absorption intensities at around 660 nm and 500 nm. The Bp^{-} disappeared with the half-life of 0.2 µs, which was about 5 times shorter than that of the radical anion of the corresponding homopolymer. On the other hand, the absorption due to Py^{-} first increased and then decreased slowly at about 0.2 µs after the pulse. Thus the existence of the intramolecular electron transfer in the PVB anion was demonstrated. The possibility of intermolecular electron transfer between polymers was rejected.

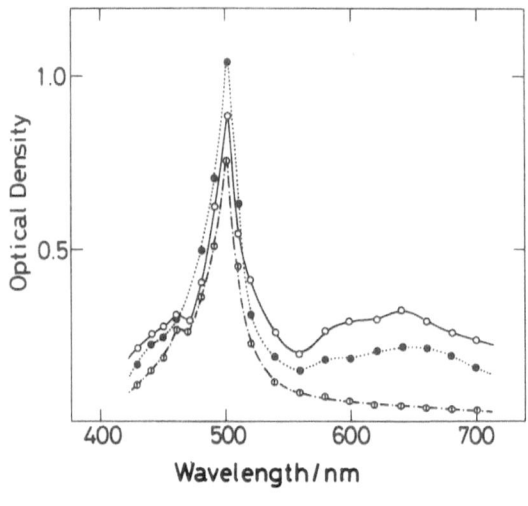

Fig. 9. Transient absorption spectra observed at ambient temperature on pulse radiolysis of MTHF solution containing 5 base-m mol dm^{-3} poly(4-vinylbiphenyl-co-1-vinylpyrene). The content of pyrenyl group is 0.15% against the total number of the side groups. (○) pulse end; (●) 0.2 μs after a pulse; (①) 1.4 μs after a pulse [60]

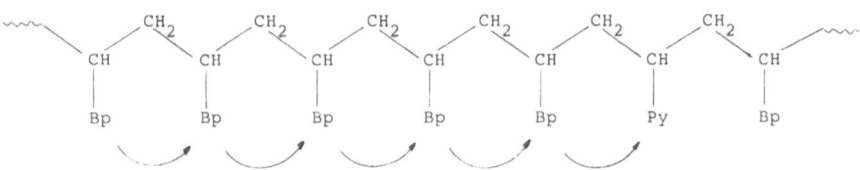

First, the spectral change was not observed with the solution containing two kinds of homopolymer at the same polymer concentrations as that in the copolymer experiments. Second, the copolymer concentration did not affect the rate of the spectral change. If the intermolecular electron transfer was actually occurring, the rate should have increased with increasing copolymer concentration.

The time-dependent spectra were analyzed on assuming one-dimensional well type potential for the copolymer at the sites where pyrene groups exist. The existence probability w of the excess electron in the region between two pyrenyl groups is expressed as a function of time t after a pulse, and the position x from one of the pyrenyl groups. The probability is obtained by solving the following partial differential equation, assuming an equal probability for the initial electron-trapping by biphenyl groups in the region.

$$\frac{\partial w(x, t)}{\partial t} = \Lambda \frac{\partial^2 w(x, t)}{\partial x^2}, \quad w(x, 0) = 1 \tag{11}$$

Where, Λ is a constant with the dimension $[\mathbf{L}]^2[\mathbf{T}]^{-1}$ which is equivalent to a one-dimensional diffusion coefficient of the electrons. The total existence probability in the section $0 < x < a$ at time t is obtained by the integration of $w(x, t)$. The distance a of two pyrenyl groups must have a certain distribution as a result of random-copolymerization: Tanaka et al. assumed the most probable distribution for it [61].

$$f(a) = p^a(1 - p) \tag{12}$$

Here, p is the content of biphenyl group in the copolymer. Thus the following formula was obtained for the total existence probability, $P(t)$.

$$P(t) = \frac{\sum\limits_{a=1}^{\infty} (8a/\pi^2) f(a) \sum\limits_{n=1}^{\infty} \{1/(2n+1)^2 \exp[-\{(2n+1)\pi\}^2 \Lambda t/a^2]}{\sum\limits_{a=1}^{\infty} (8a/\pi^2) f(a) \sum\limits_{n=1}^{\infty} \{1/(2n+1)^2\}} \qquad (13)$$

$P(t)$ is proportional to the concentration of $Bp^{\bar{\;}}$, while $1 - P(t)$ is proportional to the concentration of $Py^{\bar{\;}}$. The former decreases and the latter increases with time.

Exact fittings to the experimental curves, i.e. the time-profiles of the absorption intensity of the anions, were difficult because of the overlapping decay reaction of the polymer anions, but the upper limit of Λ was estimated as 4×10^{-10} m^2 s^{-1}. This value is 1–2 order smaller than that of the transfer coefficient of singlet energy migration in polystyrene [62] and poly(2-vinyl naphthalene) [63] via the Forster mechanism. From this value the average traveling distance of the electron for 1 μs is roughly estimated as 300 Å by using the equation $\bar{r}^2 = 2\Lambda t$.

From the similar measurements by using 1,2-dichloroethane as solvent, the migration rate of positive charge from group to group along the polymer chains was estimated. A positive charge was localized not on a single chromophore but over two chromophores to give an absorption spectra characteristic to dimer cation. Because of this interaction, the rate of the positive charge transfer is very slow in comparison with that of negative charge.

4.3 Excited States in Polymer Solutions

Transfer of energy from ionizing radiation to the solution produces various kinds of excited state. The energy transfer processes in irradiated polymer solutions can be probed by observing the excited states of the polymers.

In cyclohexane solutions, electrons produced by ionizing radiation are quasi-free electrons with a high electron mobility. They recombine with parent cations within picoseconds to yield excited states of solvent molecules. Solute excited states are produced by the energy transfer to solute molecules, although the direct excitation by low-energy secondary electrons should not be overlooked [64]. Indeed both monomer and excimer fluorescence were observed in the solutions of polystyrene in cyclohexane [65]. A strong absorption of intramolecular excimer was also observed for polystyrene (λ_{max} = 520 nm) and poly(α-methylstyrene) (λ_{max} = 520 nm) solutions [65]. The assignment is consistent with a theoretical treatment of Vala et al.: the excimer with an interplanar separation of 0.33 nm of the benzene rings was predicted to yield an energy of 2.39 eV (λ_{max} = 518 nm) for the transition $^1E_{1u} \rightarrow {}^1B_{1g}$ [66]. The lifetime of the excimer fluorescence of chloromethylated polystyrene was 20 ns, without being affected by the chloromethylation ratio of the polymers from 0 to 24% [59].

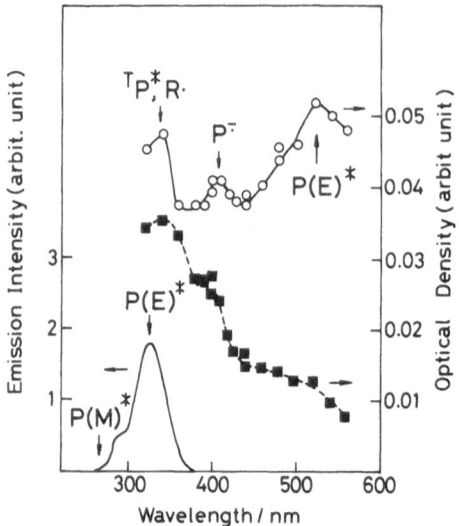

Fig. 10. Transient absorption spectra observed at (○) 2 ns and (■) 50 ns after an electron pulse in polystyrene films. Fluorescence spectrum observed in polystyrene film excited by the light of 253.7 nm (——). P(M)*: monomer fluorescence of polystyrene; P(E)*: excimer flurorescence and absorption of polystyrene; T_p^*, absorption of the excited triplet state of polystyrene; R·, absorption of radicals; $P^{\bar{\cdot}}$, absorption of the radical anion of polystyrene [67]

This value agrees with the lifetime of the excimer of polystyrene. The intensity of the excimer fluorescence decreased with increasing chloromethylation ratio. Electrons produced in cyclohexane, one of the precursors of the excimer, were scavenged by chloromethylated part of polystyrene. Absorption spectra of the excited states and the polymer radicals were measured by laser photolysis of the cyclohexane solutions. The results are summarized in Fig. 10 [67].

Itagaki et al. studied in detail the intramolecular excimer formation, singlet energy migration, and relaxation of the internal rotation of polystyrene in cyclohexane [62]. They used oligostyrene as model compound and measured transient decay-curves of monomer and excimer emissions. The rate constants were determined for the excimer formation with the numbers of monomer unit in the oligomers from 2 to 13. The rate constants for the excimer formation, of the order of 10^8 s^{-1}, increased with increasing number of monomer unit for $2 < n < 8$, then leveled off at the value for polystyrene. The rate constant for the singlet energy migration between adjacent benzene rings in polystyrene was also determined to be 3×10^{10} s^{-1}. An average time of 7.2 ns was necessary for two adjacent benzene rings to attain the excimer conformation from their initial equilibrium distribution. They also investigated the intramolecular excimer formation in *meso* and racemic diastereoisomeric dimer, 6,8-diphenyltridecanes, [68]. The difference in the rate for excimer formation of these dimers was ascribed to the conformational change for *meso* and racemic dimers.

4.4 Radicals and Charge Transfer Radical Complexes of Polystyrene and Related Polymers

Polystyrene and related polymers have most extensively been studied in organic solutions in connection with the radiation-induced degradation [57, 58, 64]. In dioxane and cyclohexane, excited states of monomers as well as intramolecular excimers were produced; polymers were hardly decomposed in these solutions. In chloroform and carbon tetrachloride, the charge transfer radical complex between polystyrene and chlorine atom was observed. The absorption spectrum of this complex had maxima at 320 nm and 500 nm (Fig. 11) and decayed according to first order kinetics with a lifetime of 400 ± 20 ns. Washio et al. have proposed the following mechanism for the formation and reaction of the complex in chloroform [57, 58]:

$$CCl_4 \rightsquigarrow CCl_4^+ + e^-$$

$$CCl_4^+ + PSt \rightarrow CCl_4 + PSt^{\ddagger}$$

$$CCl_4 + e^- \rightarrow {}^{\cdot}CCl_3 + Cl^-$$

$$PSt^{\ddagger} + Cl^- \rightarrow (PSt^{\delta+} Cl^{\delta-})$$

$$(PSt^{\delta+} Cl^{\delta-}) \rightarrow PSt^{\cdot} + HCl \tag{14}$$

where PSt and $(PSt^{\delta+} Cl^{\delta-})$ denote polystyrene and the charge transfer radical complex. As the polymers were effectively decomposed in these solutions, the complex should be a precursor of polymer radicals. Addition of cyclohexane reduced both the initial yield of the complex and the G-value of the main-chain scission of polystyrene; the square of the initial yield was correlated to the G-value of the main-chain scission. As the concentration of the complex was proportional to that of polystyrene radical, Tabata et al. presumed that the degradation proceeded through the second-order recombination reaction of the radicals [58]. In the presence of O_2, all of the polymer radicals were converted to peroxyradicals, leading to the main-chain scission of the polymers.

In benzene and toluene solutions, electrons produced by ionizing radiation also recombined with parent cations very quickly. Excited states are mostly originated through this reaction. The transient absorption obtained in the solution of chloromethylated polystyrene showed a band due to benzyl type

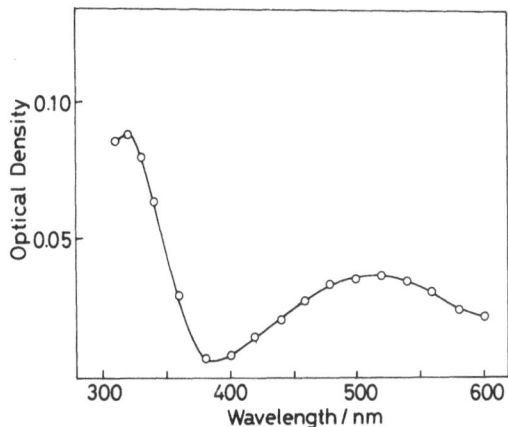

Fig. 11. Transient absorption spectrum observed at 170 ns after a pulse in 200 base-mmol dm^{-3} polystyrene in CHCl$_3$ [58]

radicals as well as that of the charge transfer radical complex between benzene molecules and chlorine atom [58, 59]. The benzyl type radical was considered to be produced by the following energy transfer from an excited benzene molecule to chloromethylated part of the polymers.

$$(15)$$

Owing to this type of reaction, both chloromethylated polystyrene and poly(α-methylstyrene) show crosslinking properties.

In carbon tetrachloride poly(α-methylstyrene) were degraded, even without oxygen, by the irradiation of ionizing radiation [58]. The decay of the charge transfer radical complex, observed in this solvent, may be due to the reaction of a chlorine atom with β-site H of poly(α-methylstyrene); the reaction leads to the formation of β-position radicals of poly(α-methylstyrene). The produced polymer radicals were unstable and dissociated into neutral and radical species.

4.5 Degradation and Crosslinking of Polymers in Solution Studied by LSI

The kinetics of degradation and crosslinking have been studied directly by following the time dependence of light scattering intensity (LSI) of the polymer solution after the electron-pulse irradiation. The LSI increases upon cross-linking or folding, whereas it decreases upon the main-chain scission or unfolding (Fig. 12). If U denotes signal voltage which is proportional to the light scattering intensity, the degree of the main-chain scission can be expressed as

$$\frac{U_0 - U}{U - U_L} = \frac{mG(S)D_{abs}}{10^2 N} \tag{16}$$

provided the radiation-induced LSI change is purely due to the main-chain scission [69]. Where the subscript 0 refers the state of the system before the pulse and L stands for solvent. Other symbols are: m is molecular weight of the repeating unit, $G(S)$ is the number of the main-chain scission, D_{abs} is absorbed dose(eV/g) and N is Avogadro's number.

The ratio $(U_0 - U)/(U - U_L)$ observed from the solutions of polyisobutene in various hydrocarbons is proportional to the absorbed dose as expected from Eq. 16 [70]. Schnabel et al. found that the LSI decreased with time in the

[Reaction]   [Change of LSI]

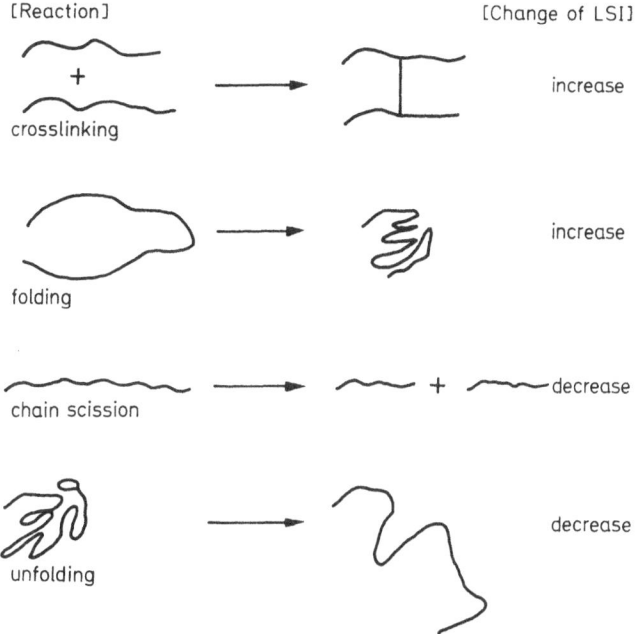

crosslinking   increase

folding   increase

chain scission   decrease

unfolding   decrease

Fig. 12. Processes leading to changes in the light scattering intensity (LSI) [74]

microsecond time-domain owing to a diminution of the size of the polymers [70]. The process of main-chain rupture consists of a series of consecutive reactions: (1) absorption of energy from ionizing radiation, (2) main-chain scission by chemical reaction, and (3) separation of fragments or disentanglement diffusion. The time-dependent LSI gives the information of the rate determining step of the process. The curves obtained from the solutions of polyisobutene were analyzed as being correlated solely with disentanglement diffusion, since the lifetime of the decay was proportional to the solvent viscosity.

In the case of PMMA dissolved in acetone, the change of LSI could be correlated both with the separation of fragments and, to some extent, with the lifetime of the intermediates which contribute to the main-chain scission [71]. The LSI decreased in two modes, probably due to the two pathways for the main-chain scission. The fast mode with a lifetime of about 20 μs was influenced neither in its extent nor in its rate by the addition of O_2 or mercaptane. Therefore the first mode was ascribed to the diffusional separation of fragments which are generated by the main-chain scission through the direct decomposition of electronically excited or ionic intermediates. The slow mode with a lifetime of 6 ms was suppressed, to an extent, depending on the O_2-concentration; it was attributed to long-lived polymer radicals. The added O_2 reacts with lateral polymer radicals to prevent their decomposition.

A peculiar LSI change has been observed upon investigating poly(methyl vinyl ketone) [72]. As shown in Fig. 13, the initial decrease with a lifetime of 20 μs in the LSI after the pulse was followed by a slow increase. The increasing LSI indicated the existence of the augmentation of the averaged molecular weight due to intermolecular crosslinking. The rate of the LSI decrease was not influenced by the presence of O_2 in accordance with the prediction that the main-chain breaks occurred very rapidly, e.g. via Norrish type II processes. It was also consistent with the explanation that the rapid decrease in the LSI was ascribed to the separation of the fragments of the polymers produced by the scission of the C–C bond in the main-chain. On the other hand, the rise of the LSI due to the crosslinking was prevented by O_2 completely as demonstrated by the trace (b) in Fig. 13. The inhibition of crosslinking was interpreted in terms of O_2 reacting with lateral polymer radicals according to the following reactions.

$$R^{\cdot} + O_2 \; \rightarrow \; R—O—O^{\cdot} \tag{17}$$

It was shown in the pulse radiolysis of the aqueous solution of poly(ethylene oxide), for example, the peroxy radicals produced by the reaction of O_2 combined and formed highly unstable oxyl radicals [73]. The LSI decay-curve after the pulse observed with an O_2-saturated solution showed two modes. The faster one obeyed a second order kinetics, suggesting that Eq. (17) was the rate determining step in the series of consecutive reactions. This reaction was followed by H-abstraction of OH radical, leading to the main-chain scission.

$$R + {}^{\cdot}OH \; \rightarrow \; R^{\cdot} + H_2O \tag{18}$$

$$R^{\cdot} + O_2 \; \rightarrow \; R—O—O^{\cdot} \tag{19}$$

$$ROO^{\cdot} + {}^{\cdot}OOR \; \rightarrow \; R—O—O—O—O—R \tag{20}$$

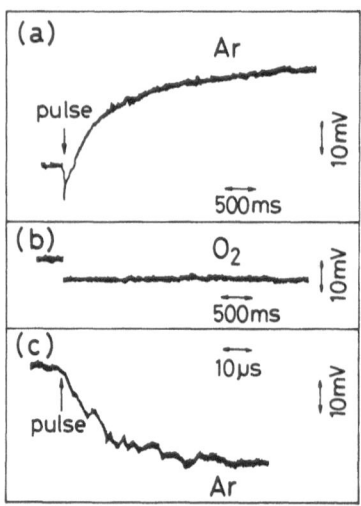

Fig. 13a–c. Main-chain degradation and intermolecular crosslinking of poly(methyl vinyl ketone) in acetone solution following the irradiation at ambient temperature with a 2 μs pulse. Changes in the LSI: in Ar-saturated solution (a) and (c), and in O_2-saturated solution (b) [74]

$$R\!-\!O\!-\!O\!-\!O\!-\!O\!-\!R \left[\begin{array}{l} \rightarrow 2RO^{\cdot} + O_2 \rightarrow \text{fragment radicals} + O_2 \quad (21a) \\[2ex] \underset{\displaystyle R'\!-\!C\!=\!O}{\overset{\displaystyle R''}{}} + \underset{\displaystyle R'\!-\!\underset{\displaystyle H}{\overset{\displaystyle R''}{C}}\!-\!OH}{} + O_2 \quad (21b) \end{array} \right.$$

As the $G(S)$ was found to equal to $G(OH)$, the average number of radical site per initial polymer can easily exceed unity at absorption dose of few krad. Under these experimental conditions, the fractions of peroxy radicals may undergo intramolecular instead of intermolecular reactions.

As exemplified in the reactions of PMMA and poly(ethylene oxide), the role of the O_2 in the reactions of polymer radicals is quite important. Schnabel classified the effect of O_2 on the oxidative degradation processes in linear polymers as follows: (1) O_2 acting as a promoter of main-chain degradation, (2) O_2 acting as inhibitor of main-chain degradation, and (3) O_2 acting as a fixing agent for main-chain breaks [74].

4.6 Reaction of Polymers and Biological Materials in Aqueous Solution

Earlier pulse radiolysis investigations were devoted to the radiation effects of the polymers dissolved in water. The rate constants of the reactions of OH radicals with poly(ethylene oxide) [75, 76], polyvinylpyrolidone [75], dextran [75], and sodium polyacrylate [77] have been measured as functions of the chainlength and the polymer concentrations. The rate constant expressed in polymer unit $(\text{mol}^{-1}\,\text{dm}^3\,\text{s}^{-1})$ increased more slowly with increasing chainlength than would be expected if the reactivity of the CH_2 and CH groups were additive. The reaction between small molecules and the polymer radicals produced by the reaction of OH radical with poly(ethylene oxide) has also been demonstrated [78].

Braams and Ebert found in the early studies on enzymes that the reaction of a hydrated electron e_{aq}^- with ribonuclease was affected by the conformation of the polymers: the rate constant increased abruptly upon unfolding of the polymers as the temperature increased [79]. The reaction of OH radical with catalase was diffusion-controlled although part of the amino acids of catalase did not react at every collision with an OH radical under isolated conditions [80]. Lindenau et al. studied the degradation of native DNA by the time dependency of LSI. Electron irradiation induced the decrease of the LSI consisting of fast and slow modes [81]. The fast LSI decrease was due to the separation of segments generated by the single strand scissions located directly opposite to each other. The slow mode, on the other hand, was due to the detachment of the segments generated by single strand breaks at sites on alternate strand which was separated by several nucleotide units. The LSI increase after the pulse was also observed in the absence of O_2, indicating the occurrence of the crosslinking reaction. The LSI increase was permanent at relatively low doses, but it was completely suppressed by O_2 [82].

In the pulse radiolysis of aqueous solutions of poly(styrenesulfonate), various radical species have been identified [83]. OH radicals reacted with this polymer to produce a mixtures of OH adducts. When several OH radical adducts were produced on the same polymer molecule, the intramolecular radical-radical reactions occurred. A radical cation was also produced from the OH adduct by the following acid catalyzed reaction.

$$-CH-CH_2- \qquad\qquad -CH-CH_2-$$

The radical cation decayed in the time scale of hours: the reaction was attributed to an intramolecular conversion of the radical cation into a benzyl type radical, which subsequently decayed to produce stable recombination products.

Numerous experiments have been made by pulse radiolysis on the reduction reaction of e_{aq}^- in biological and biochemical systems. Comprehensive survey on this subject is beyond the scope of this review. But, for example, the pulse radiolysis study on the reduction of earthworm hemoglobin is worth mentioning, because it has relevance to the degradation of macromolecules in solutions [84]. This hemoglobin consisted of 10 subunits with the total molecular weight as large as 4 000 000, the central heme being embedded at the position 5 nm inside from the surface. The e_{aq}^- decayed out within 2 μs after the pulse, whereas it took 20 μs to complete the reduction of Fe^{3+} to Fe^{2+}. This time-lag may suggest relatively slow migration of the electron from the surface to the heme through the protein. Time-dependent LSI measurements indicated that the dissociation of the subunits occurred in the time scale of 100 μs (Fig. 14).

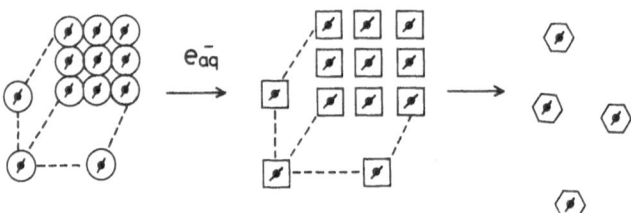

Fig. 14. A schematic model of the dissociation of earthworm hemoglobin induced by the reduction by hydrated electron. The molecule consists of 10 subunits with the total molecular weight of 4 000 000 [84]

5 Rigid and Molten Polymers

5.1 General Remarks

In principal, the primary processes induced by ionizing radiation in rigid polymers are the same as those in fluid solutions: ionization and excitation are followed by the formation of new reaction intermediates such as ions, excited states, and radicals. However, the reactions of these intermediates may be much affected by the chemical and physical structure of the medium polymer. Besides, a large variety of chemical additives in the polymers make it difficult to understand the early reactions induced by ionizing radiation.

Because of the experimental difficulties, pulse radiolysis had not been applied to solid polymers for a long time, except for few cases [85, 86]. Recently, however, owing to the development of the experimental techniques as well as the swelling interest in industry, various rigid polymers including polystyrene [87], PMMA, chloromethylated polystyrene, ethylene-propyrene-diene-terpolymer [88], and epoxy resin [89] were investigated. Most of the experiments were carried out with doped materials; additive-free polymers were not frequently used, because the absorbance change induced by a single pulse in neat polymers was not so large and the observed spectra were usually so broad and uncertain.

5.2 Polyethylene and Related Compounds

5.2.1 Alkanes

The pulse radiolysis studies of liquid alkanes have relevance to the radiolysis of polyethylene and related polymers. In liquid alkanes at ambient temperature, the reaction intermediates such as alkane radical-cations, olefin radical-cations, olefine dimer-cations, excited states, and alkyl radicals have been observed after the electron-pulse irradiation [90–93]. According to the nanosecond and sub-nanosecond studies by Tagawa et al., the observed species were alkane radical cations, excited states, and alkyl radicals in n-dodecane; excited states and cyclohexyl radical were observed in cyclohexane, and only radicals in neopentane [91, 93]. Olefin radical-cations were also detected in cyclohexane containing carbon tetrachloride [92].

The alkane radical-cations generated in electron-pulse irradiated n-dodecane show an absorption band in the visible with its maximum at 800 nm (Fig. 15) [93]. The position of the absorption maximum changed from 600 nm to 900 nm depending on the carbon number of the alkane. It was noted that the lifetime of alkane-radical cations was shorter than that of the solvated electrons observed in the near infrared region. These phenomena were interpreted in terms of the following ion-molecular reaction.

$$RH_2^+ + RH_2 \ \rightarrow \ RH + RH_3^+ \tag{22}$$

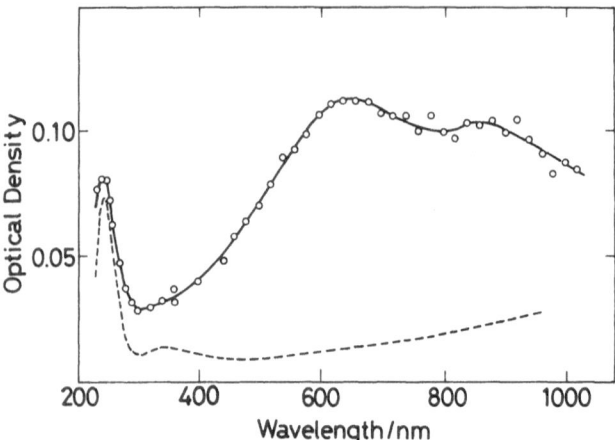

Fig. 15. Transient absorption spectra observed at ambient temperature in pulse radiolysis of neat liquid n-dodecane (n-$C_{12}H_{26}$) immediately ($\bigcirc$) and at 80 ns (- - - -) after a 2 ns electron pulse irradiation [93]

Another broad absorption in the visible, which was observed in the shorter wavelength side of the cation band, was assigned to $S_n \rightarrow S_1$ transition; its lifetime coincided with that of the fluorescence [93]. The absorption maximum also shifted to the longer wavelength side with increasing carbon number of alkanes, but it tended to saturate when the carbon number exceeded twenty. In the lower alkanes, the distinction between the excited-state band and the cation band was not clear.

The assignment of the transient visible-absorptions induced in cyclohexane are still in dispute. First, Trifunac et al. claimed that the short-lived absorption at 600 nm, observed on nanosecond time-scale, was due to a positive ion [92]. The assignment seems to be made on the analogy of the results obtained by n-hexane [90]. However, Tagawa et al. concluded from the lifetime and the effect of scavengers that the very broad and structureless visible band should be due to the excited states of cyclohexane [93]. On the other hand, Mehnert et al. decomposed the transient visible-absorption, which was generated in cyclohexane containing 5 mmol dm^{-3} CCl_4, into two different cationic transients [94]. They assigned the short-lived transient peaking at about 500 nm to the cyclohexane radical-cation. However, in contrast to the case of n-alkanes, the radical cation should be very short-lived in cyclohexane and difficult to be detected by conventional nanosecond pulse radiolysis.

A relatively sharp absorption in the UV region due to alkyl radicals is observed in electron-pulse irradiated alkanes [93]. It has an absorption maximum at 240 nm in n-dodecane and cyclohexane. (Mehnart et al. did not see this absorption maximum but found another short-lived absorption band peaking at 270 nm in n-hexane, n-heptane, and n-hexadecane containing 10 mmol dm^{-3} CCl_4. This absorption band was assigned to olefin monomer radical-cation produced by the fragmentation of alkane radical-cation [94].) The formation and reaction of alkyl radical in irradiated alkane are interesting in conjunction with the mechanism of radiation-induced decomposition of polyethylene. The time-profile of alkyl radical observed at 240 nm after a 20 ps pulse in cyclohexane showed a prompt formation of the radical within the pulse duration [93]. It means that the formation of the radical is synchronized neither with the decay of the excited singlet nor the one of the radical cation. Consequently, neither the reaction (22) nor the following two reactions can be main sources of alkyl radicals.

$$RH_2^* \rightarrow \ ^\cdot RH + H^\cdot \qquad\qquad\qquad (23)$$

$$RH_2 + H^\cdot \rightarrow \ ^\cdot RH + H_2 \qquad\qquad\qquad (24)$$

Recent synchrotron radiation experiments showed that the probability of the energy deposition on the alkane molecules was the highest at about 16–18 eV [95]. With the energies above 16 eV, excited states of alkane radical cations can be produced efficiently. In irradiated cyclohexane, for example, the following reactions were considered to be the formation reactions of alkyl radical.

$$RH_2^{+*}(RH_2^{+}) \rightarrow RH^+ + H^*(H^\cdot)$$

$$RH^+ + e^- \rightarrow \ ^\cdot RH^*$$

$$^\cdot RH^* \rightarrow \ ^\cdot RH$$

$$^\cdot RH^* \rightarrow \ R + H^\cdot \qquad\qquad\qquad (25)$$

Tagawa et al. concluded that the alkyl radicals were produced mainly from excited states of radical cations (RH_2^{+*}) and partially from higher excited states of cyclohexane (RH_2^{**}) and other sub-ordinate reactions [93].

5.2.2 Polyethylene and Ethylene-Propylene Copolymer Films

Figure 16 shows the absorption spectrum obtained by additive-free polyethylene [67]. At ambient temperature the absorption observed on nanosecond time-scale increased continuously from 500 to 200 nm without showing any maximum. The absorption in UV is similar to that obtained by γ-irradiation. Considering the results obtained by liquid alkanes, the absorption seems to be comprised of several different free radicals. At 95 K additional absorption due to the trapped electron was observed at wavelengths longer than 600 nm; the band was observable even at ambient temperature in the picosecond time-domain [96]. The electron decays presumably by the hole-electron recombination. The decay of the trapped electron was independent of the presence of carbon tetrachloride, suggesting that the additives reacted with a mobile electron but not with the trapped electron. On adding naphthalene, the radiation-induced spectrum showed the bands due to the first excited triplet state and the radical

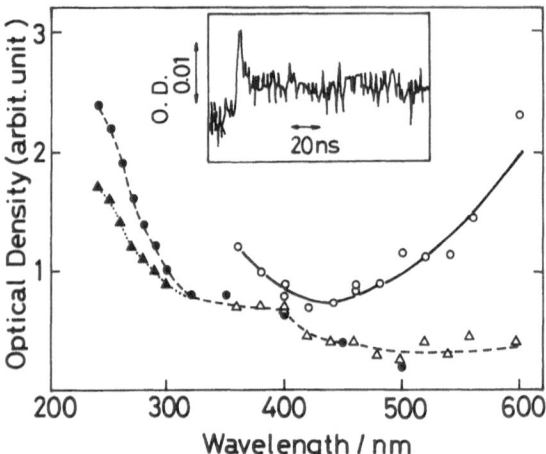

Fig. 16. Transient absorption spectra observed after a pule at (●) 100 ns and (▲) 10 µs in polyethylene film and at (○) 2 ns and (△) 50 ns in ethylene-propylene rubber [67]

anion of naphthalene [97]. With pyrene as solute, the absorptions of the solute radical-anion and radical-cation were observed together with that of the triplet states. The transient decayed only partially leaving the remaining absorption practically constant in the microsecond time-domain. The emission spectrum from pyrene-doped sample showed a series of sharp peaks partially due to pyrene-excimer fluorescence. As the excimer emission was observed immediately after the pulse at pyrene concentrations as low as 10^{-3} mol dm^{-3}, the adjacent parallel pyrene molecules were supposed to exist with a large overlap such as was found in the pyrene crystal.

Johnson and Willson interpreted the main feature of the observations on solid polyethylene doped with aromatic solutes in terms of an ionic mechanism; it was analogous to that proposed for irradiated frozen glassy-alkane-systems in which ionization occurred with G = 3 − 4 [96]. The produced charged species, electron and positive hole, were both mobile as indicated by the radiation-induced conductivity. The production of excited states of aromatic solutes was caused mainly by ion-electron neutralization. The ion-ion recombination was relatively slow but it might contribute to the delayed fluorescence observed. On the basis of Debye-Simoluchovski equation, they evaluated the diffusion coefficients of the radical anion of naphthalene and pyrene as approximately 4×10^{-12} and 1×10^{-12} m^2 s^{-1} respectively: the values were about three orders of magnitude less than those found in typical liquid systems.

Ethylene-propylene copolymer films gave a very broad absorption in the visible region upon electron-pulse irradiation (Fig. 16) [93]. It was comprised of at least three species, electrons, excited states, and alkane radical cations. At about 700 nm and 800 nm the contributions from excited states and radical cations, respectively, were largest. The lifetime of the radical cation determined at about 800 nm was less than 10 ns. The radical cation was supposed to decay quickly by the proton transfer reaction from the analogy of liquid alkanes.

5.2.3 Molten Polyethylene

Several problems arise when solid polymers are chosen as objects of pulse radiolysis experiment. Some polymers scatter the analyzing light beam because of the semicrystallinity of the matrices. The radiolysis products accumulated in the sample cause uncertainties in the identification of the observed transients. To overcome these problems, Brede et al. used pure and blended polyethylene in the molten and transparent state [98–103]. They pressed the polymers into 10 mm thick plates and then prepared the samples by cutting them into pieces. The samples were adjusted on a block-type irradiation cell with unprotected windows for electron entrance and analyzing light path. The samples were heated through the block electrically up to 393 K just before the pulse radiolysis measurements. The results obtained with these samples have provided interesting and meaningful information on the chemical reactions induced by ionizing radiation in polyethylene.

As observed in liquid alkanes, a relatively long-lived absorption ($\tau_{1/2}$ = 500 ns) in UV and a short-lived ($\tau_{1/2} \leq 10 \ \mu s$) one in the visible region were induced by electron-pulse irradiation in molten polyethylene [102]. From the effects of scavengers, it was obvious that the short-lived absorption in the visible region was due to cationic species. Brede et al. ascribed it to the olefine radical-cation; it may be formed by the fragmentation of the primary parent polyethylene-cation or the migration of the positive hole through the polymer chain and its trapping by unsaturated groups in the polymer [102]. The cationic species decayed probably by the neutralization reaction. Electrons, a candidate of reaction partners, were very short-lived and could not be detected on nanosecond time-scale. The decay of the olefine radical-cation was accelerated by the addition of electron scavenger; thus the reaction partner of the neutralization reaction may be a negative charge which is stabilized, for example, in a form of chloride ion.

The absorption induced in UV was assigned to various kinds of alkyl radical as was the case of liquid alkanes. The absorption decayed according to a first-order law, but a very long-lasting background remained. The decaying and the remaining components were altogether interpreted as due to the allyl radical formation via the self-decomposition of the alkyl radicals [101].

$$-\underset{\underset{H}{|}}{\overset{\overset{H}{|}}{C}}-\underset{\underset{H}{|}}{\overset{\overset{H}{|}}{C}}-\underset{\underset{H}{|}}{\overset{}{\dot{C}}}- \quad \rightarrow \quad H_2 + -\underset{\underset{H}{|}}{\overset{}{C}}{=}\underset{\underset{H}{|}}{\overset{}{C}}-\underset{\underset{H}{|}}{\overset{}{\dot{C}}}-. \qquad (26)$$

The formation of a branched polyethylene radical with primary radical function or a tertiary radical structure was also considered in the molten polyethylene saturated with ethylene.

$$-CH_2-\overset{\cdot}{C}H-CH_2- + C_2H_4 \;\rightarrow\; -CH_2-CH-CH_2-$$
$$\qquad\qquad\qquad\qquad\qquad\qquad\quad \overset{|}{C}H_2$$
$$\qquad\qquad\qquad\qquad\qquad\qquad\quad \overset{|}{\cdot}CH_2$$

$$\rightarrow\; -CH_2-\overset{\cdot}{C}-CH_2- \qquad\qquad (27)$$
$$\qquad\qquad\quad \overset{|}{C}H_2$$
$$\qquad\qquad\quad \overset{|}{C}H_3$$

Another interesting phenomenon observed was the unusually fast formation rate of the scavenger radical. The diffusion-controlled reactions in this system can not be so fast because of the high viscosity of the molten polyethylene at 393 K; it was expected to occur in the microsecond time-domain. Actually, except for few cases, the scavenger radical were formed within the time-resolution of the experiments of 40 ns. From these observations Brede et al. have proposed an excitation migration model, i.e. a rapid motion of the excitation along the polymer chain with energy transfer to the foreign molecules [102]. The vibrational motions of the similar molecules are so strongly coupled that the whole system is supposed to vibrate as a single molecule. In such a system the excitation may be spread over the whole of molecules; a traveling wave packet moves through the system in a coherent manner with constant group velocity. This model may be applied to the radical formations in irradiated liquid alkanes too.

5.3 Polystyrene and Related Polymers

Radiation-induced reactions in polystyrene and poly(α-methylstyrene) has drawn much attention; they are frequently used in industry as electron-beam negative(crosslinking) and positive(scission) resists, respectively [104]. Many years ago, Ho and Siegel observed the absorption spectrum of the polystyrene radical-anion with a maximum at 419 nm in additive free polystyrene [86]. The fast initial decay of the absorption was ascribed to the recombination of ion pairs formed in close proximity. The recent pulse radiolysis studies by Tagawa et al. demonstrated additional two bands at 530 and 1000 nm due to the excimer of polystyrene and dimer radical-cation of benzene, respectively [59]. It means that both electrons and positive holes should migrate in the solid polymers and eventually be trapped in some trapping sites to form the polystyrene radical-anion and the dimer radical-cation. Excited singlet states also migrate through the polymer chains in the distance of about 8 monomer units and become excimers and triplet state. Washio et al. have observed the absorption spectra of the excited triplet states and radicals [105].

Chloromethylated polystyrene and chloromethylated poly(α-methylstyrene) are negative type resists having high sensitivity and high resolution. In the pulse radiolysis of solid films of this polymer, the absorption spectra of substituted benzyl-type polymer-radical and the charge transfer complex between phenyl rings and chlorine atoms were observed (Fig. 17) [59]. The benzyl-type radical may be produced by the dissociative electron attachment to the benzyl part of chloromethylated polystyrene (CMS).

$$e^- + CMS \rightarrow P_1^* + Cl^- \tag{28}$$

The chloride ion formed by this reaction interacts with a CMS cation to form the charge transfer complex, which in turn becomes the precursor of the following two types of polymer radicals (P_2^*):

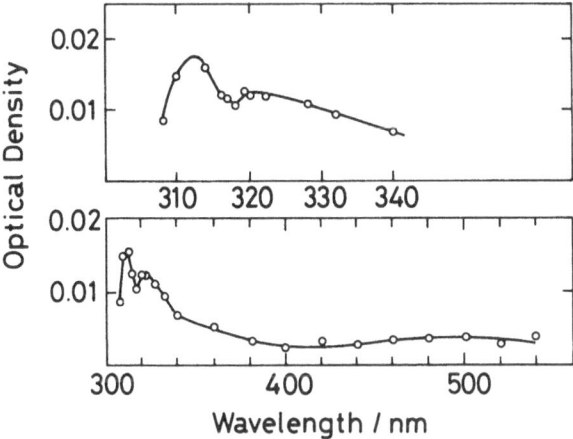

where R is CH_2Cl or H. These radicals should be produced very near to P_1^* and the crosslinking will take place between P_1^* and P_2^*.

Very recently Kouchi et al. constructed an ion beam pulse radiolysis system and use it for the study of the LET effect in irradiated polystyrene thin films [106]. The nanosecond pulsed MeV ion beam with the variable repetition rate was obtained by chopping ion beams from a Van de Graaff. Time profiles of the excimer fluorescence from polystyrene thin films, excited by He^+ impact, were

Fig. 17. Transient absorption spectra observed in chloromethylated polystyrene solid films [59]

recorded by a photon counting technique with a coincidence method. The low dose-time profile (the irradiation time of 0–139 s) showed a single exponential decay, with the lifetime similar to that obtained by the electron pulse radiolysis study. However, the high dose-time profile (the irradiation time of 1839–3839 s) seemed to consist of two components, the fast decay component and the slow one. The lifetime of the slow decay component was in agreement with that by electron pulse radiolysis. The fast decay component was ascribed to the quenching of excimers by products at the region overlapped by two tracks formed by separately incident ions.

5.4 Poly(Methyl methacrylate)

With additive-free PMMA, Ho and Siegel observed a broad, structureless absorption in the region from 350 to 500 nm, with the radiation yield of $G\varepsilon = 5000 \, mol^{-1} \, dm^3 \, cm^{-1} \, (100 \, eV)^{-1}$ at 370 nm and $2600 \, mol^{-1} \, dm^3 \, cm^{-1}$ $(100 \, eV)^{-1}$ at 500 nm [86]. However, assignment of the absorptions was not possible at that time, because nothing was known about reaction intermediates produced in PMMA by radiolysis. Recently, Tabata et al. observed short-lived absorptions peaking at 725 nm $(G\varepsilon = 3000 \, mol^{-1} \, dm^3 \, cm^{-1} \, (100 \, eV)^{-1}$, $\tau_{1/2} = 190$ ns at $-13°C$) and at 440 nm; the absorptions were ascribed to the cation and the anion of the polymer, respectively [107]. However, these assignments are not compatible with the results obtained by Ogasawara et al. in the solutions of PMMA and its substituted analogues [46]. Very recent experiments showed that the 440 nm band was due to PMMA radical-anion and the 725 nm band was probably due to the cationic species derived from acetone remaining in the samples [108].

In doped PMMA, solute ions were efficiently formed on exposure to ionizing radiation: cations were formed for most cases, but anions were formed only when the solute had a high electron affinity [86]. Kira et al. irradiated solid PMMA containing excess biphenyl and a small amount of a second solute by electron-pulses and observed the absorption spectrum of the biphenyl radical-cation produced by the following reactions [109]:

$$PMMA^{\dot{+}} + Bp \rightarrow Bp^{\dot{+}} \tag{29}$$

where $PMMA^{\dot{+}}$ denotes a positive hole in a PMMA matrix whose nature is not clearly understood. This reaction was followed by the positive-charge transfer to the second solute with a lower ionization potential. In this experiment no absorption corresponding to biphenyl radical-anion was detected. This is consistent with the previous steady-state measurements on γ-irradiated PMMA film by Torikai et al. [110], but apparently conflicting with the results by the classical work by Ho and Siegel [86]: the absorption spectrum shown in that paper was composed of only the triplet state and the biphenyl radical-anion. We do not know the reason for this discrepancy.

5.5 Effect of Additives and Hardeners
on the Radiation Resistance

Addition of aromatic compounds effectively increases the radiation resistance of polymers. Perhaps the added aromatics affect the subsequent cascade of energy deactivation paths to release the excitation energy. The bond cleavage will also be affected in the presence of the additives. Although the mechanism of the radiation effect must be quite complex, the pathways of transfer of radiation-induced excitation energy from polymer matrix to aromatic molecules could be probed by observing time-dependent emission spectra. A few pioneering works along this line have already been reported [88, 89].

Kawanishi et al. studied the effect of aromatic additives on the radiation resistance of ethylene-propylene-diene-terpolymer by picosecond pulse radiolysis [88]. Fluorescence of acenaphthene and excimer emission of acenaphthylene were observed in terpolymer containing acenaphthene and acenaphthylene, respectively. When dicumyl peroxide was used as a crosslinking agent instead of di-t-butyl peroxide, another emission band ascribed to the reaction product of dicumyl peroxide and naphthalene ring of the aromatic additives was observed. The weaker monomer fluorescence observed in the acenaphthylene system was due to smaller quantum efficiency of the excited state. The acenaphthylene is grafted onto polymer chain in a form of oligomer as shown in Fig. 18, which facilitates nonradiative transfer from polymer to acenaphthene. Stronger excimer emission observed in the acenaphthylene system was attributed to the difference in dispersion of the aromatics in polymer matrix: acenaphthene molecules were well dispersed without forming coagulation, whereas acenaphthylene was dispersed in coagulated forms in which excimers could easily be formed at low concentrations of additives. Coagulation may occur by grafting as shown in Fig. 18.

Effects of aromatic additives on the radiation resistance may be summarized as followed. Addition of acenaphthylene or acenaphthene provides an effective

a) At system

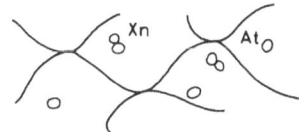

b) Ay system

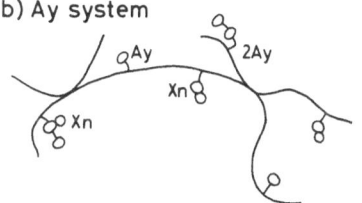

Fig. 18a, b. A schematic model of the dispersion of the aromatic additives in polymer matrix. **a)** Acenaphthene system: acenaphthene (At) and reaction products between aromatic fragment radicals from crosslinking agent and the naphthalene ring of the additives (Xn) dispersed in polymer chains. **b)** Acenaphthylene system: acenaphthylene (Ay) and Xn molecules are fixed to polymer chains [88]

route for release of the excitation energy in the form of fluorescence. When an aromatic peroxide such as dicumyl peroxide is used as crosslinking agent, a new route for energy release formed in the system enhances radiation stability of the matrix polymer. Grafted oligomer chains of acenaphthylene contribute to the excimer formation as well as the excitation energy release in the form of heat through a nonradiative transition.

The mechanisms of radiation damage and effects of hardeners were studied recently by pulse radiolysis [89]. The epoxy resins require a relatively large amount of curing agents (hardeners), most of them are aromatic and aliphatic amines such as diamino diphenyl methane or triethylene tetramine. On the basis of the emission spectra and kinetic behavior of excited states observed, the radiation resistance of aromatic and aliphatic amine curing epoxy resin was explained by internal radiation protection effects due to energy transfer.

6 Conclusion

Many of the pulse radiolysis studies, especially in the early days, were made in aiming at characterizing reaction intermediates in the initial stages of the radiation-induced polymerizations. However detection of intermediates by this method is in principle limited to the most abundant species which occur at a concentration high enough to produce measurable signal intensities. One can speculate that the observed transient species may be connected to the main route of the reactions, but this is not always true. For instance, most of the polymerization experiments suggested the cationic mechanism for the radiation-induced polymerization of styrene, but only radical-anions and neutral radicals were detected in earlier pulse radiolysis studies; the formation of cationic initiators was not demonstrated. This is because the G value of the initiators of cationic polymerization is only one tenth of the initial G value of ion-pair production and, besides, the monomeric radical-cations of styrene were consumed so efficiently to form dimer radical-cations, leading to the propagating cationic ends. In this case, the important intermediates were not observed just because they participated in the main route of the polymerization.

Despite of this inherent limitation, several spectacular results have been obtained. It should be noted that the initiation mechanism of the cationic polymerization of styrene described above was also deduced from the results of pulse radiolysis experiments. The pulse radiolysis combined with other techniques, such as the matrix isolation technique, the electron spin resonance technique and usual polymerization techniques, definitely provides a powerful means for investigating fundamentals of polymerization.

Several authors have applied pulse radiolysis to the reactions of the polymers with some reactive intermediates, such as solvated electrons and OH radicals in solutions. The reactions were found to be affected by the molecular weight, the conformational change, and the presence or absence of entanglement of polymers. The experiments on radiation-induced degradations and cross linking of polymers by use of time-dependent light scattering have also revealed unique features of the polymer reactions.

Pulse radiolysis experiments on solid polymers have provided new insight into the mechanism of radiation damage of polymers. Recent studies on some practically important polymers clarified the pathways of transfer of radiation-induced excitation energy from polymer matrix to additives; thus the roles of additives in the radiation resistance or sensitivities of polymers are understood in terms of elementary energy transfer processes. The usefulness of this method is verified not only in the basic science but also in the field of application.

Acknowledgment. The author is grateful to Prof. S. Tagawa of the University of Tokyo and Prof. H. Kobayashi of Tsukuba University for their helpful discussion and careful reading of the manuscript.

7 References

1. Fessenden RW, Schuler RH (1963) J Chem Phys 39: 2147
2. Schmidt WF, Allen AO (1970) J Chem Phys 52: 4788
3. Pikaev AK (1967) Pulse radiolysis of water and aqueous solutions. Indiana University Press, Bloomington
4. Hunt JW, Thomas JK (1967) Radiat Res 32: 149
5. Bronskill MJ, Taylor WB, Wolff RK, Hunt JW (1970) Rev Sci Instr 41: 333
6. Tagawa S, Katsumura Y, Tabata Y (1979) Chem Phys Letters 64: 258
7. Kobayashi H, Ueda T, Kobayashi T, Washio M, Tabata Y (1983) Radiat Phys Chem 21: 13
8. Kobayashi H, Ueda T, Kobayashi T, Tagawa S, Yoshida Y, Tabata Y (1984) Radiat Phys Chem 23: 393
9. Williams F (1968) Principles of radiation-induced polymerization. In: Ausloos P (ed) Fundamental processes in radiation chemistry. Wiley Interscience, New York
10. Katayama M, Hatada M, Hirota K, Yamazaki H, Ozawa Y (1965) Bull Chem Soc Jpn 38: 851
11. Katayama M (1965) Bull Chem Soc Jpn 38: 2208
12. Katayama M, Yamazaki H, Ozawa Y, Hatada M, Hirota K (1966) J Chem Soc Jpn Pure Chem Sec (Nippon Kagaku Zassi) 87: 37
13. Schneider C, Swallow AJ (1966) J Polym Sci Polym Letters 4: 277
14. Metz DJ, Potter RC, Thomas JK (1968) J Polym Sci A-1, 5: 877
15. Ronayne MR, Guarino JP, Hamill WH (1962) J Am Chem Soc 84: 4230
16. Keene JP, Land EJ, Swallow AJ (1965) J Amer Chem Soc 87: 5284
17. Chambers KW, Collinson E, Dainton FS, Seddon WA, Wilkinson F (1967) Trans Faraday Soc 63: 1699
18. Schneider C, Swallow AJ (1968) Makromol Chem 114: 155
19. Ueno K, Hayashi K, Okamura S (1966) Polymer 7: 431 and references therein
20. Chapiro A (1974) Makromol Chem 175: 1181
21. Swallow AJ (1968) Adv Chem Ser No.82, 499
22. Yoshida H, Noda M, Irie M (1971) Polymer J 2: 359
23. Mehnert R, Brede O, Nauman W (1982) Ber Bunsenges Phys Chem 86: 525
24. Egusa S, Arai S, Kira A, Imamura M, Tabata Y (1977) Radiat Phys Chem 9: 419
25. Egusa S, Tabata Y, Arai S, Kira A, Imamura M (1978) J Polym Sci Polym Chem Ed 16: 729
26. Shida T, Hamill WH (1966) J Chem Phys 44: 4372
27. Badger B, Brocklehurst B (1969) Trans Faraday Soc 65: 2582
28. Hayashi K, Irie M, Lindenau D, Schnabel W (1978) Radiat Phys Chem 11: 139
29. Nauman W, Müller E, Brede O, Mehnert R (1981) Radiochem Radioanal Letters 47: 391

30. Brede O, Mehnert R, Nauman W (1982) Radiat Phys Chem 20: 155
31. Rzad SJ, Infeha PP, Warman JM, Schuler RH (1970) J Chem Phys 52: 3971
32. Silverman J, Tagawa S, Kobayashi H, Katsumura Y, Washio M, Tabata Y (1983) Radiat Phys Chem 22: 1039
33. Langan, JR, Salmon G (1983) J Chem Soc Faraday Trans I 79: 589
34. Szwarc M (1968) Carbanions, living polymers and electron-transfer processes. Wiley-Interscience, New York, p 368
35. Ogasawara M, Kajimoto N, Izumida T, Kotani K (1985) J Phys Chem 89: 1403
36. Mah S, Yamamoto Y, Hayashi K (1982) J Polym Sci Polym Chem Ed 20: 1709
37. Mah S, Yamamoto Y, Hayashi K (1982) J Polym Sci Polym Chem Ed 20: 2158
38. Mah S, Yamamoto Y, Hayashi K (1984) Radiat Phys Chem 23: 137
39. Tagawa S, Arai S, Imamura M, Tabata Y, Oshima K (1974) Macromolecules 7: 262
40. Tagawa S, Tabata Y, Arai S, Imamura M (1974) J Polym Sci Polym Letters Ed 12: 545
41. Washio W, Tagawa S, Tabata Y (1980) J Phys Chem 84: 2876
42. Tagawa S, Arai S, Kira A, Imamura M, Tabata Y, Oshima K (1972) J Polym Sci Polym Letters Ed 10: 295
43. Tabata Y (1976) J Polym Sci Symp No. 56, 409
44. Arai S, Kira A, Imamura M (1977) J Phys Chem 81: 110
45. Ogasawara M, Arai S, Imamura M (1979) J Polym Sci Polym Letters Ed 17: 649
46. Ogasawara M, Tanaka M, Yoshida H (1987) J Phys Chem 91: 937
47. Tanaka M, Yoshida H, Ogasawara M (1988) Radiat Phys Chem 32: 719
48. Matsushima M, Kato N, Miyazaki T, Fueki K (1987) Radiat Phys Chem 29: 231
49. Tanaka M, Ogasawara M, Yoshida H (1990) Radiat Phys Chem 36: 243
50. Kato N, Miyazaki T, Fueki K, Yokoi T (1989) Macromolecules 22: 4124
51. Ban H, Sukegawa K, Tagawa S (1987) Macromolecules 20: 1775
52. Ogasawara M, Suganuma T, Junke N, Yamaoka H, Yoshida H (1992) Radiat Phys Chem 40: 111
53. Tachikawa H, Yoshida H, Ogasawara M (1991) Radiat Phys Chem 37: 107
54. Nelson JT, Pietro WJ (1988) J Phys Chem 92: 1365
55. Bobrowski K, Das PK (1987) J Phys Chem 91: 1210
56. Irie S, Irie M (1986) Macromolecules 19: 2182
57. Washio M, Tagawa S, Tabata Y (1983) Radiat Phys Chem 21: 239
58. Tabata Y, Tagawa S, Washio M, Hayashi N (1985) Radiat Phys Chem 25: 305
59. Tagawa S (1986) Radiat Phys Chem 27: 455
60. Tanaka M, Yoshida H, Ogasawara M (1991) J Phys Chem 95: 955
61. Ham G (1964) Copolymerization. Wiley-Interscience, New York, p 8
62. Itagaki H, Horie K, Mita I, Washio M, Tagawa S, Tabata Y (1983) J Chem Phys 79: 3996
63. Itoh S, Yamamoto M, Nishijima Y (1981) Polym J 13: 791
64. Thomas JK, Johnson K, Klippert T, Lowers J (1968) J Chem Phys 48: 1608
65. Tagawa S, Schnabel W, Washio M, Tabata Y (1981) Radiat Phys Chem 18: 1087
66. Vala Jr MT, Hillier IH, Rice SA, Jortner J (1966) J Chem Phys 44: 23
67. Yamaoka H, Tagawa S (1984) J At Energy Soc Japan 26: 739
68. Itagaki H, Horie K, Mita I, Washio M, Tagawa S, Tabata Y, Stato H, Tanaka Y (1985) Chem Phys Letters 120: 547
69. Beck G, Kiwi J, Lindenau D, Schnabel W (1974) Eur Polym J 10: 1069
70. Beck G, Lindenau D, Schnabel W (1975) Eur Polym J 11: 761
71. Beck G, Lindenau D, Schnabel W (1977) Macromolecules 10: 135
72. Lindenau D, Beavan SW, Schnabel W (1977) Eur Polym J 13: 819
73. Gröllmann U, Schnabel W (1980) Makromol Chem 181: 1215
74. Schnabel W (1986) Radiat Phys Chem 28: 303
75. Behzadi A, Borgwardt U, Henglein A, Schamberg E, Schnabel W (1970) Ber Bunsenges Phys Chem 74: 649
76. Matheson M, Mamou A, Silverman J, Rabani J (1973) J: Phys Chem 77: 2420
77. Behzadi A, Schnabel W (1973) Macromolecules 6: 824
78. Schnabel W (1978) Hoshasen Kagaku 26: 23
79. Braams R, Ebert M (1968) Adv Chem Series 81: 464
80. Henglein A, Karman W, Robke W, Beck G (1966) Makromol Chem 92: 105
81. Lindenau D, Henglein A, Schnabel W (1976) Z Naturforsh 31C: 484
82. Washino K, Schnabel W (1982) Makromol Chem 183: 697
83. Behar D, Rabani J (1988) J Phys Chem 92: 5288

84. Tamura M, Hayashi K, Lindenau D, Schnabel W, unpublished data
85. Ho SK, Siegel S, Schwarz HA (1967) J Phys Chem 71: 4527
86. Ho SK, Siegel S (1969) J Chem Phys 50: 1142
87. Tagawa S (1983) Proc Int'l Ion Engineering Congress 1681
88. Kawanishi S, Hagiwara M, Katsumura Y, Tabata Y, Tagawa S (1985) Radiat Phys Chem 26: 707
89. Tagawa S, Washio M, Hayashi N, Tabata Y (1985) J Nucl Mat 133 & 134: 785
90. Jonah CD (1983) Radiat Phys Chem 21: 53
91. Tagawa S, Washio M, Kobayashi H, Katsumura Y, Tabata Y (1983) Radiat Phys Chem 21: 45
92. Trifunac AD, Sauer Jr MC, Jonah CD (1985) Chem Phys Letters 113: 316
93. Tagawa S, Hayashi N, Yoshida Y, Washio M, Tabata Y (1989) Radiat Phys Chem 34: 503
94. Mehnert R, Brede O, Cserep G (1985) Radiat Phys Chem 26: 353
95. Koizumi H, Yoshimi T, Shinsaka K, Ukai M, Morita M, Hatano Y, Yagishita A, Itoh K (1985) J Chem Phys 82: 4856
96. Yamaoka H, Tagawa S (1984) J At Energy Soc Japan 26: 739
97. Johnson GRA, Willson A (1977) Radiat Phys Chem 10: 89
98. Brede O, Hermann R, Wojnarovits L, Stephan L, Taplick T (1987) Zfl-Mitt Leipzig Nr 132: 101
99. Brede O, Stephan L (1987) Zfl-Mitt Leipzig Nr 132: 217
100. Brede O, Hermann R, Helmstreit W, Taplick T, Stephen L (1988) Makromol Chem Macromol Symp 18: 113
101. Brede O, Naumann W (1988) Radiat Phys Chem 32: 475
102. Brede O, Hermann R, Wojnarovits L, Stephen L, Taplick T (1989) Radiat Phys Chem 34: 403
103. Brede O, Luther B (1989) Isotopenpraxis 25: 6
104. Imamura S (1979) J Electrochem Soc 126: 1628
105. Washio M, Tagawa S, Hayashi N, Tabata Y (1984) Polymer Preprints Jpn 33: 279
106. Kouchi N, Aoki Y, Shibata H, Tagawa S, Kobayashi H, Tabata Y (1989) Radiat Phys Chem 34: 759
107. Tabata M, Nilsson G, Lund A, Shoma J (1983) J Polym Sci Polym Chem Ed 21: 3257
108. Ogasawara M, unpublished result
109. Kira A, Imamura M (1984) J Phys Chem 88: 1865
110. Torikai A, Kato H, Kuri Z (1976) J Polym Sci Polym Chem Ed 14: 1065

Editor: S. Okamura
Received December 11, 1991

Radiation Synthesis of Polymeric Materials for Biomedical and Biochemical Applications

I. Kaetsu

Department of Nuclear Reactor Engineering, Faculty of Science and Technology, Kinki University, Kowakae 3-4-1, Higashi-Osaka 577, Japan

The study of Radiation Induced Polymerization in the supercooled state is reviewed. This polymerization has remarkable characteristics owing to a rapid increase of viscosity, such as a large polymerization rate at low temperatures and a maximum rate at temperatures above the glass transition temperature (T_g).

Applications of polymerization in a supercooled state to the immobilization of various biofunctional components is reviewed. Those applications show advantages because in the low temperature biofunctional components such as proteins, drugs and cells are entrapped or adhered effectively in the polymerized matrix. The immobilized composites are used for biomedical and biochemical systems and processes, such as immuno-diagnosis, artificial organs, drug delivery systems and cell cultures.

1 Introduction

Radiation can contribute to biotechnology and bioengineering by various chemical and physical means, such as polymerization, grafting, crosslinking and etching. This research field can be divided into two categories: the synthesis of polymers having a special biological function, and the preparation of polymeric composites including a special biological component by radiation immobilization. Research in this field can also be classified into two areas according to the purposes of the application: the biomedical applications such as the synthesis of biocompatible polymers and the immobilization of drugs for drug delivery systems, and the biochemical applications such as immobilization of biocatalysts and cells for various bioreactors and analytical systems.

Various investigations have been carried out in these categories and application areas. The industrial applications are not wide spread as yet, but this

Advances in Polymer Science, Vol. 105
© Springer-Verlag Berlin Heidelberg 1993

research is increasingly attracting the interest of researchers as a new and promising field. The author has studied radiation-induced polymerization in the supercooled state at low temperatures. After basic and systematic research into this polymerization, the applications to the immobilization of various biofunctional components such as enzymes, antibodies, drugs, organelles, microorganisms and tissue cells were carried out. It was found that the radiation polymerization in the supercooled state gave unique immobilization techniques and the successful results. In this chapter, the studies on the polymeric materials, especially the immobilized composites for biomedical and biochemical applications of radiation polymerization are reviewed.

2 Polymerization in the Supercooled State and the Application to Immobilization

The polymerization in the supercooled state or glassy state was initially of interest to several workers [1–6], but has not received as much attentions as that shown in crystalline state polymerization. Kaetsu et al. found that many acrylates and methacrylates monomers could be supercooled and remain stable [7–9] and have studied this polymerization field extensively and systematically since 1967 [10–13]. The most remarkable characteristic of the supercooled monomer is the sharp and sudden increase of viscosity with the temperature decrease. The viscosity of the supercooled monomer increases exponentially and

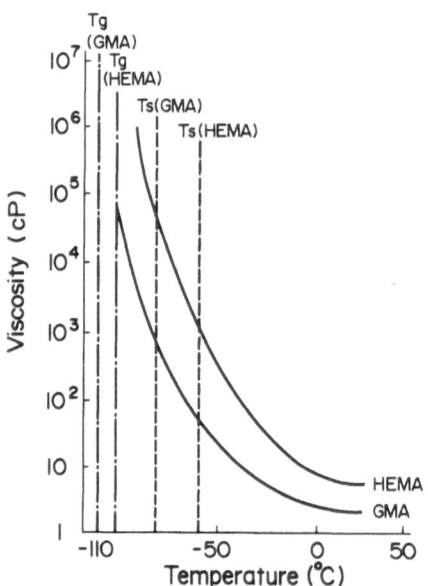

Fig. 1. Viscosity change of supercooled monomers as a function of temperature. Monomer: HEMA (2-hydroxyethyl methacrylate). GMA (glycidyl methacrylate). T_g: glass transition temperature. T_s: temperature of maximum polymerization rate

transforms to the glassy state on reaching the glass transition temperature (T_g). No in-source polymerization occurs at temperatures below T_g, but rapid post-polymerization takes place when the irradiated monomer is warmed up to temperatures above T_g. This is so because monomer mobility is retarded and the radicals formed are trapped at temperatures below T_g. The polymerization rate in an in-source polymerization above T_g is large and has a maximum at a certain temperature (T_s). The conditions and the amount of maximum polymerization rates are changed by various polymerization factors and conditions.

A saturated conversion was observed when the monomer-polymer mixture reached a composition, in which the T_g was the same as the polymerization temperature. However, as almost all supercooled monomers have T_g-values between $-50\,°C - -150\,°C$, it is easy to obtain a 100% conversion for most

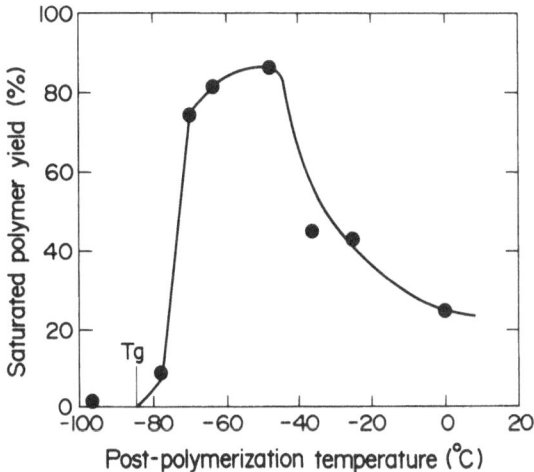

Fig. 2. Saturated polymer yield as a function of polymerization temperature in post-polymerization of supercooled monomer. Monomer: acrylonitrile-succinic acid-acetamide. Irradition: 4×10^5 rad·h⁻¹, $-196\,°C$. Post-polymerization time: 1 h

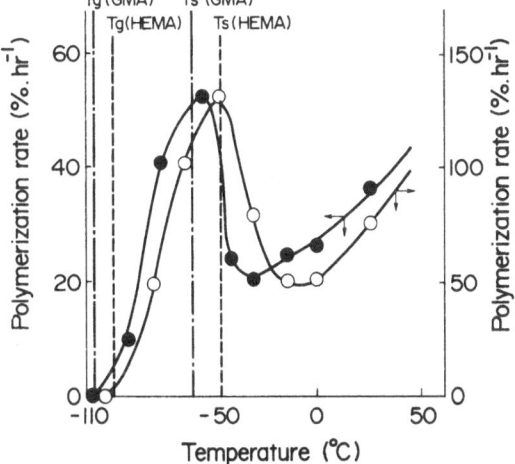

Fig. 3. Temperature dependency of polymerization rate in supercooled monomers. Monomer: ○ HEMA (2-hydroxyethyl methacrylate); ● GMA (glycidyl methacrylate). Irradiation: 1×10^5 rad·h⁻¹

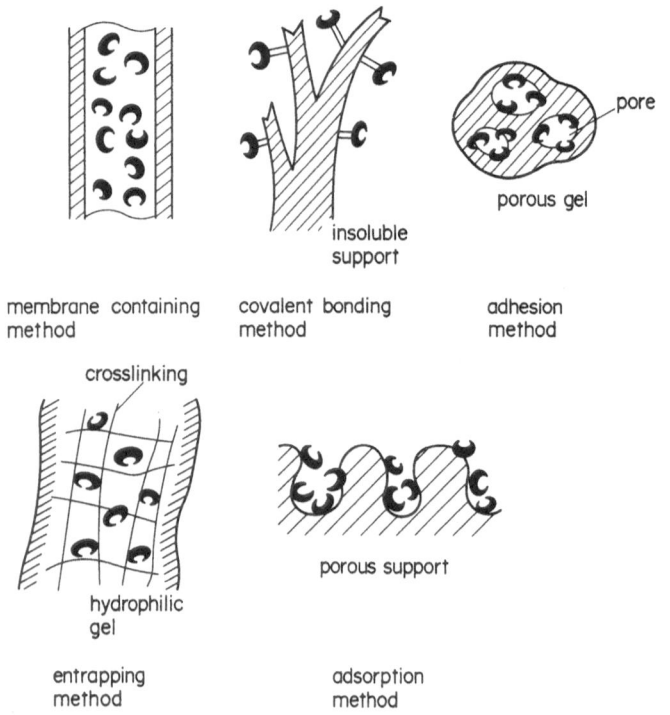

Fig. 4. Model scheme for the products of main immobilization methods. **c**: biofunctional component

Table 1. Various factors affecting the degree of immobilization in the physical entrapping method

Factors	Direction of control	Fixation	Release
Factors in biocomponent			
Molecular weight	Increase	Promote	Retard
Concentration	Increase	Retard	Promote
Solubility to medium	Increase	Retard	Promote
Homogeneity of dispersion	Increase	Retard	Promote
Factors involved during application			
Water content	Increase	Retard	Promote
Temperature	Increase	Retard	Promote
pH	Increase	Retard or Promote	Promote or Retard
Factors in polymer			
Hydrophilicity	Increase	Retard	Promote
Degradability	Increase	Retard	Promote
Rigidness	Increase	Retard	Promote
Porosity	Increase	Promote	Retard
Surface area	Increase	Retard	Promote
Adsorbent	Increase	Promote	Retard
Multi-layer structure	Increase	Promote	Retard

systems. The T_g of a monomer can be raised by increasing the pressure in the system. Polymerization in the supercooled state was first applied to a casting process for organic glass-materials for optical uses [14–16]. It was found that this polymerization process could control the temperature-rise due to the polymerization heat and reduce the volume contraction. Thus, polymeric lenses of high optical quality and excellent surface smoothness and dimensions were obtained in a very short time cycle. This process was used and developed practically for the light focusing lenses by the Nikon Company, Japan. The second application was the immobilization of biofunctional components. Immobilization can be defined as including a biofunctional component in a limited small volume. There have been a large number of immobilization methods reported in the literature since the 1960's. These can be classified into four types: inclusion in a membrane-formed container, adsorption on a solid support, entrapping or adhesion in a polymeric gel, and chemical binding with a polymeric support-material. The purposes of immobilization are; continuous or repeated use of biofunctional components, shaping into an relevant form and

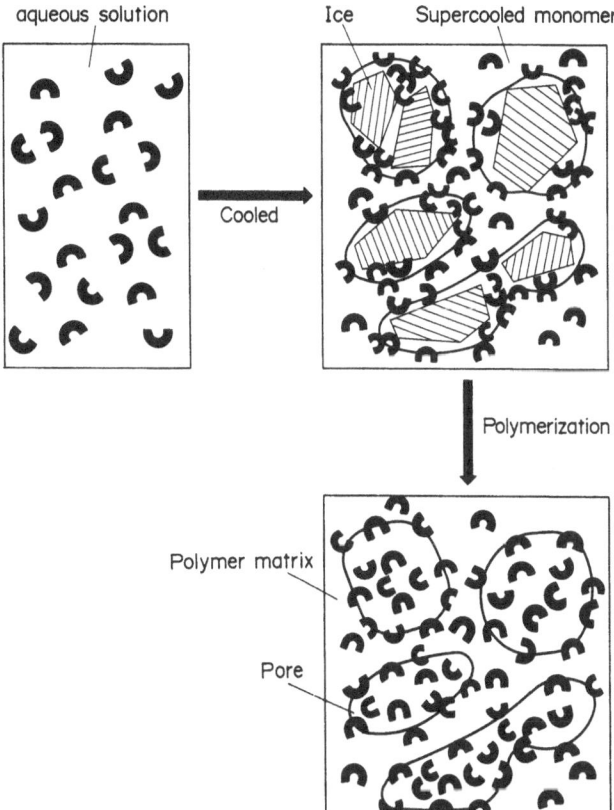

Fig. 5. Model scheme for the immobilization process by radiation polymerization in a supercooled state. (Adhesion method) **c**: biofunctional component

size for operation, increasing stability and extention of optimum conditions for practical uses, enhancing of cell cultures, and controlled and sustained release of the entrapped component. This immobilization is now becoming an important technology for biomedical devices and biochemical reactors.

Among several immobilization methods, a physical entrapment or adhesion has the advantage of wide applicability to almost all kinds of biofunctional components. It can be used for cell immobilization with cell growth and also for drug immobilization for controlled release, because all kinds of entrapped biofunctional components can move inside the polymeric matrix. We can adjust the three variables – fixation, release, and growth of component – by changing the polymeric matrix. The product immobilized by radiation polymerization in the supercooled state was studied using a composite of a physically entrapped biofunctional component with polymethacrylate or polyacrylate. The super-cooled state has two main advantages for the immobilization. These are: prevention of radiation damage of the biofunctional component owing to the freezing of water to reduce the attack of radicals from water against biofunc-tional molecules, and increase of porosity (inner surface area) in the polymeric matrix due to the freezing of water to form a porous structure after melting. As the result, the activity of the immobilized composite increased remarkably in the supercooled temperature region. The results of the applications of radiation-immobilized composites to protein fixation, drug release, and cell culture will be reviewed in the following sections.

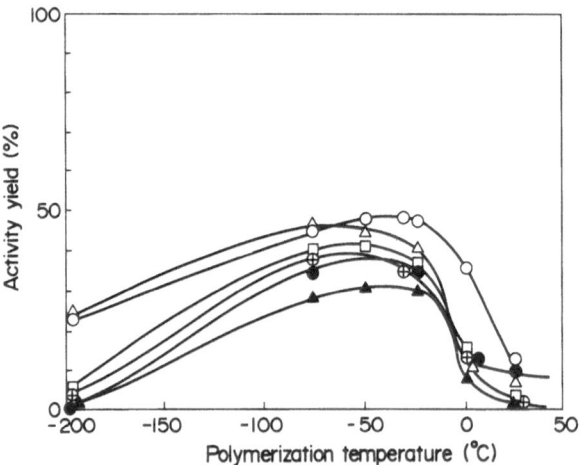

Fig. 6. Temperature dependency of activity yield in the immobilization of enzymes by radiation polymerization. Enzyme: ○ α-amylase in 50% HEMA; △ glucoamylase in 50% HEMA; □ cellulase in 50% HEMA; ⊕ glucose oxidase in 50% HEMA; ▲ glucoamylase in 30% HEMA; ● α-glucosidase in 50% HEMA. Monomer: HEMA (2-hydroxyethyl methacrylate). Irradiation. 1×10^6 rad, in vacuo. % monomer: in buffer solution

3 Immobilization of Proteins and Applications to Analytical Systems

Dobo was the first to report an immobilization of an enzyme with radiation [17]. Immobilization of enzymes by gamma-ray polymerization and electron beam crosslinking were studied by Kawashima [18] and Maeda [19], respectively. Kaetsu et al. have studied the immobilization of various biofunctional components by gamma-ray polymerization of supercooled acrylates and methacrylates [20–25].

In this polymerization, the biofunctional component (enzyme) can be concentrated in an interfacial area between the frozen ice crystal and the supercooled monomer phase, and immobilized by molecular entanglement between the enzyme and polymer molecules. This is a different procedure for fixation from the usual entrapping method with a crosslinked structure in a gel. Therefore, we may call this procedure the adhesion-method to distinguish it from the usual entrapping. This term was extended to cover the use of the usual synthetic polymers including hydrophobic polymers as the supports. One of the characteristic properties of products obtained in this way was that there is a maximum activity at a certain monomer concentration. The maximum activity is observed when the increased inner surface area is balanced by the increased leakage of enzyme and these occur with a decrease of monomer concentration. Immobilization by physical entrapping was also studied by Rosiak [26], Carenza [27] and Ha [28].

Another method of immobilization, using radiation graft copolymerization, has been studied by Hoffman, Gaussen, Garnnett and other researchers in the

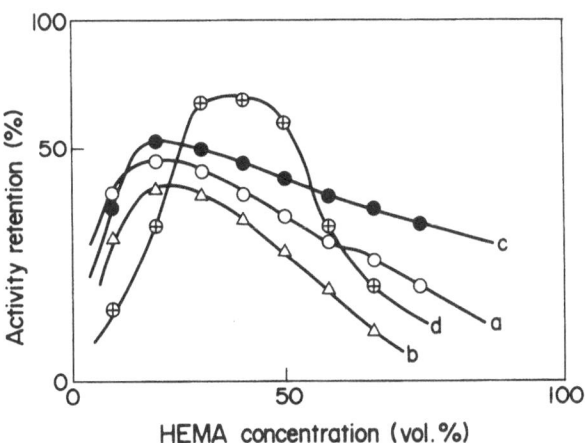

Fig. 7. Activity retention as a function of monomer concentration in the immobilized enzymes by radiation polymerization method. Enzyme: $\oplus$ α-amylase ($-24\,°C$, 1×10^6 rad); $\bullet$ glucose isomerase ($-45\,°C$, 1×10^6 rad); $\bigcirc$ α-glucosidase ($-78\,°C$, 1×10^6 rad); $\triangle$ glucoamylase ($-78\,°C$, 1×10^6 rad). Monomer: HEMA (2-hydroxyethyl methacrylate)

U.S. and Europe. Hoffman proposed an immobilization method of radiation grafting and subsequent chemical binding between the functional groups in an enzyme and the grafted polymer [29, 30]. Garnnett reported enzyme immobilization with styrene-grafted or nitrostyrene-grafted polyethylene by a chemical binding method [31]. Guthrie studied the immobilization of various enzymes with grafted copolymers such as acrylic acid-grafted nylon, hydroxyethyl acrylate-grafted cellulose and maleic anhydride-grafted starch [32]. Recently, Gao et al. have synthesized polymethacryloxy succinimide and poly-acryloxy succinimide by radiation polymerization and immobilized glucoamylase, estramustine and testosterone by chemical binding by the succinimide group of the polymers [33].

Immobilized enzymes have been applied for medical-analysis and -examination. For example, the immobilized glucoseoxidase, cholesterol oxidase, phospholipase and urease can be used for medical examinations and biosensing in various forms. Kaetsu studied the immobilization of enzymes for medical uses by radiation polymerization and grafting in the form of tubes and membranes [34]. The immobilized antibodies have been studied and applied to immunological diagnosis. Kumakura and Kaetsu developed a novel antibody immobilization technique for enzyme immuno-assay by physical entrapping using low temperature radiation polymerization [35–38]. The α-fetoprotein antibody was immobilized on the surfaces of porous thin membranes, fine particles and porous cling-film. The products showed good reactivity with the antigen and proved their practical usefulness as immunodiagnostic devices. Catt investigated an immobilization of human growth-hormone antibody for radio-immuno-assay in a disc form by chemical binding it with a modified styrene-grafted tetrafluoroethylene [39]. Rembaum studied the preparation of microspheres by radiation polymerization of hydroxyethylmethacrylate, acrylic

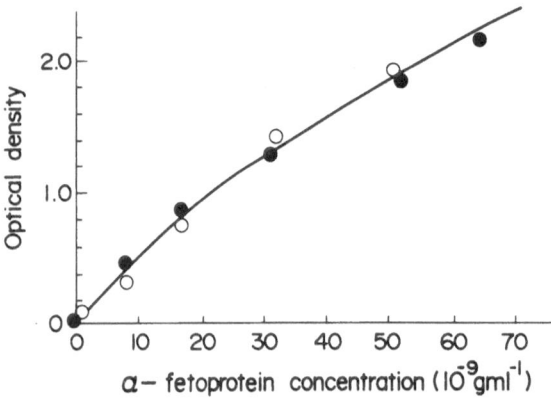

Fig. 8. Activity of immobilized antibody by the radiation polymerization method as a function of antigen concentration. Antigen: α-fetoprotein. Antibody: anti-α-fetoprotein. Immobilization method: ● polymerization of HEMA in a thin porous membrane; ○ polymerization of HEMA in particle form

acid and methacrylic acid in an emulsion and also studied the immobilization of antibodies for radioimmuno-assay, phagocytosis and cell labelling [40]. Duval and Nicaise developed an immunoadsorbant by radiation grafting of acrylic acid, acrylamide and allylamine onto polyvinyl chloride and polystyrene and immobilization of human globulin by chemical binding [41]. Leininger synthetized polycationic gels by radiation grafting of various monomers to hydrophobic supports and quaternization. Heparin was also immobilized electrostatically on the polycation for antithrombogenetic materials [42].

4 Immobilization of Drugs and Application to Drug Delivery Systems

4.1 Automatically Controlled Release Systems

Drug delivery systems have the two functions. The first is controlled drug release so as to prolong the duration of pharmacological efficacy. The other is targeting, i.e. to deliver and concentrate the drug efficiently and effectively to the desired organ or tissue. Three types of polymer drug composites are classified by the route of administration and drug delivery. These are transdermal delivery system, implantable delivery system, and injectable delivery system. The transdermal or mucosal system has a membrane or laminate form and is attached to the skin or mucous membrane. The implantable system has various molded forms such as needle, tablet and bead. It has two methods of administration. These are direct implantation in the desired organ or tissue, and subcutaneous insertion or injection followed by diffusion and circulation of the drug through the blood vessels.

Kaetsu and coworkers have developed various techniques for preparing implantable drug delivery systems by radiation polymerization, curing and crosslinking or degradation [43–45]. The fundamental methods of preparation for implantable systems can be divided into three groups. They are cast polymerization, hot-pressing, and suspension or emulsion polymerizations. The release profiles of drug can be controlled over a wide range by the design and change of various factors in the polymer such as molecular weight, copolymer composition, chemical structure, stereo-isomerism, hydrophilicity, biodegradability and porosity. The release profiles can also be varied by radiation crosslinking and degradation of the polymers in some cases. The cast polymerization method is used for the preparation of systems consisting of non-biodegradable polymers such as vinyl polymers and the hot-pressing method is used for systems consisting of biodegradable polymers. A non-biodegradable polymer-drug composite has a longer period of efficacy than a biodegradable composite. However, the interest in biodegradable polymers has increased among researchers in recent years. Various natural proteins such as collagen,

albumin and globulin can be used as biodegradable polymers. The biodegradability of those proteins can be changed by chemical or biological modification,
through heating or irradiation. The author found that the biodegradability of
globulin and albumin is decreased very much by thermal denaturation in which
the various conditions such as temperature, pressure, time and water content are
key factors [46, 47]. The biodegradation of these proteins was retarded by
irradiation. For example, the biodegradability of globulin decreased with the
increase of irradiation dose, though it increased again with further irradiation as
shown in Fig. 9. The result was explained first by the increase of crosslinking
with the initial irradiation dose and then followed predominant degradation at
the larger irradiation dose to enhance the enzymatic decomposition. In the case
of synthetic biodegradable polymers, it would be easier to design their chemical
structure and to control the biodegradability. Polypeptides or polyaminoic
acids and polyhydroxy acids are the typical biodegradable synthetic polymers.
The radiation effect on biodegradability of polypeptides is also strongly affected
by the chemical structure [48, 49]. For example, the biodegradation of poly-D,L-
alanine was affected by irradiation, while the biodegradation of copolypeptides
of D,L-alanine with other aminoic acids such as γ-ethyl glutamate and β-benzyl-
L-glutamate showed no significant effects when irradiated. Polylactic acid also
displayed a considerable effect of irradiation on its biodegradability [50].
Polydepsipeptide, which has sequential repetition of an amino acid unit and a
hydroxy acid unit, is one of the most interesting synthetic biodegradable
polymers [51]. This polymer would be a good model for the study of the effect of
chemical structure of the polymer on the biodegradability, because the molecular design of the unit and sequential composition of this polymer is possible. For
example, it was elucidated that the biodegradation was faster in the polydepsipeptide having the shorter side chain group in the aminoic acid unit.
Radiation increased the biodegradability of polydepsipeptide having the relatively shorter side chain length. The biodegradability of non-biodegradable and

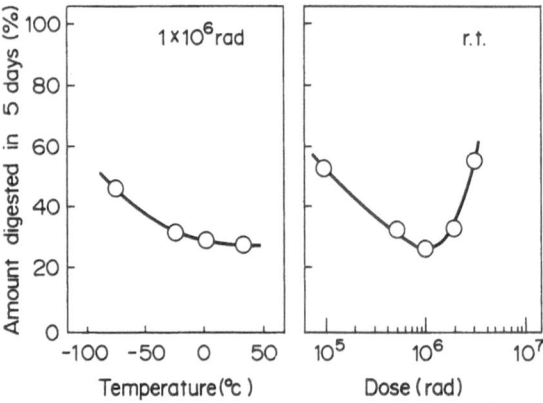

Fig. 9. Effect of irradiation on degradation of albumin as the function of irradiation temperature
and dose

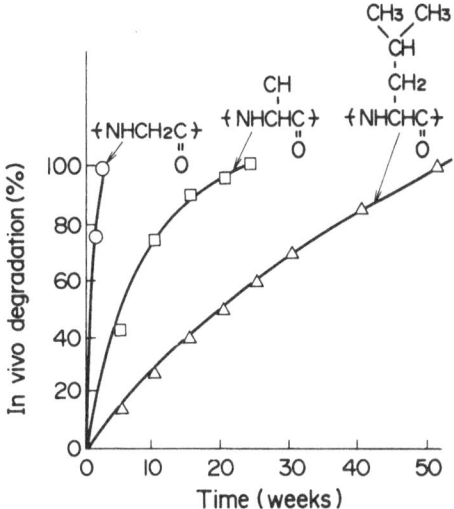

Fig. 10. Effect of side chain length in the amino acid moiety of polydepsipeptides on the in vivo degradability

biodegradable polymers mixture was changed by the composition of the two polymers [52]. Recently, Kaetsu et al. reported the results of some systematic studies on the relationship between the structure and properties of polymers and the in vivo biodegradability [53]. The implantable drug delivery systems developed by Kaetsu and coworkers have been applied to anticancer-polymer composites for local chemotherapy of cancers in digestive organs and the brain and to hormone-polymer composites for hormone-therapy of cancers in urinary organs. Those implantable composites have various forms such as needle, pellet, and button or tablet forms and have been applied in many clinical tests for cancer patients since 1977 and demonstrated successful pharmacological effects in those cancer therapies [54–57]. Kaetsu, Yamanaka, and Yoshida also

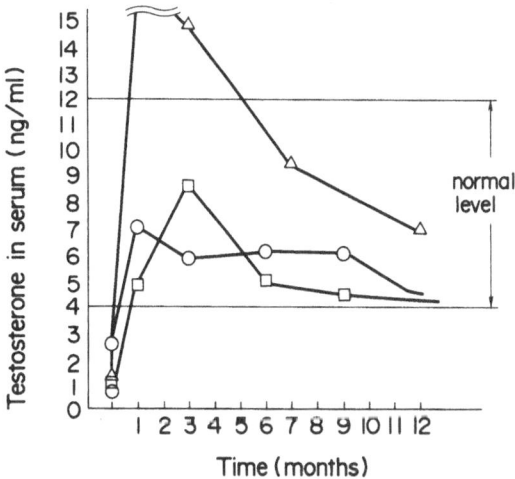

Fig. 11. Transition of serum concentration of testosterone after implantation of radiation polymerized artificial testis in clinical tests involving three people. Monomer: HEMA (2-hydroxyethyl methacrylate). Irradiation: 2×10^6 rad

developed the artificial testis by radiation polymerization of a 2-hydroxyethyl methacrylate-testosterone mixture at low temperature. 6.4 g of testosterone was included in 10 g of polymer and the product showed constant, long-lasting testosterone release. The testosterone concentration level in the blood was kept constant for one or two years in the clinical tests. Remarkable recovery of male characteristics such as body and facial hair, deep voice and body shape have been observed after the implantation of the artificial testis.

4.2 Signal-Sensitive and Responsive-Controlled Release Systems

In recent years, a very compact biosensor or a signal-recognizing device using an immobilized biofunctional component such as an enzyme, antibody, or receptor, has attracted the keen interest of researchers. On the other hand, a hydrophilic polymer having a property of reversibly changing its volume for an input of stimulation such as temperature change, pressure addition, pH change, light irradiation, ultrasonic irradiation and addition of an electric field, is becoming the subject of great interest. This type of polymer is called a stimulus-sensitive hydrogel.

Some researchers have reported the synthesis and characterization of stimulus-sensitive hydrogels. The work by Tanaka is well known [60]. Recently, Kim [61], Ishihara [62], Hoffman [63], reported the results of studies in this field. Hoffman studied the reversible volume change of hydrogels at a specific temperature which is called the lower critical solution temperature (LCST) using a radiation polymerized copolymer of isopropyl acrylamide-methylene-bis-acrylamide or methacrylic acid [63]. Kim and Okano studied the reversible volume change of polyacryloyl piperridine and polyacryloyl pyrroridine in response to the input of various stimuli such as temperature change, pH change and electric field [61]. Irie studied the photo-sensitive and responsive poly-

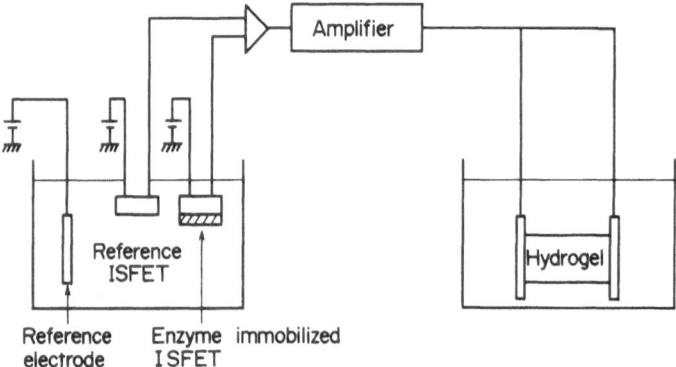

Fig. 12. Model scheme for an on-off switching drug release system consisting of a biochipsensor and a drug including an electrically responsive hydrogel

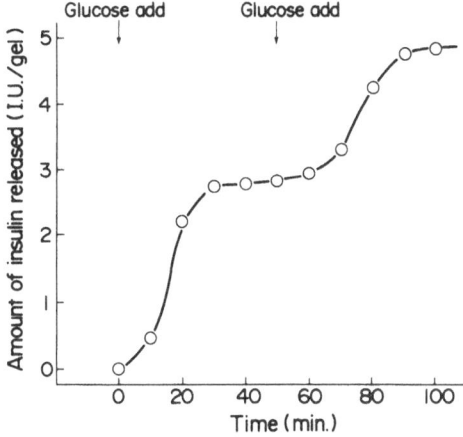

Fig. 13. Release of insulin from an on-off switching insulin release system consisting of a glucose-sensor and an N-isopropylacrylamide-acrylic acid-diethylene glycol dimethacrylate copolymer gel, with the on-off addition of glucose. (The gel is a 1 cm cube of approx. 1 g – in the test state it has absorbed water and has swollen)

acrylamide gel containing triphenylmethane leukocyanamide and tried an application to an on-off controlled release of theofilin [64].

It is considered that one of the most general and important signals for drug delivery would be a change of concentration in any kind of biochemical component in the in vivo environment. Therefore, a novel controlled-release system consisting of a biochipsensor to transduce an original signal to an electrical signal and an electrically responsive hydrogel including a drug, would be required for the purpose of a signal-responsive release system. Kaetsu and Morita developed an intelligent insulin deliver system, in which a biochip-glucose sensor (ISFET) was connected to an electrically responsive hydrogel including an insulin. The on-off switching release of insulin was observed repeatedly according to the on-off input of glucose in this new type of drug delivery system [65, 66].

5 Immobilization of Cells and Application to Cell Cultures

As already mentioned, one of the advantageous applications of physical entrapping immobilization is the cell immobilization for an enhanced cell cultures and their uses for substance production, bioconversion, biochemical reaction, hybrid-type artificial organs, and biosensors. Kaetsu, Kumakura, and Fujimura have studied cell immobilization and cell cultures using three methods involving radiation-prepared polymer supports. These methods are entrapping in a gel matrix, absorption into a porous fibrous support, and the adsorption onto a polymer surface. The first type of immobilization by gel entrapping is a suitable application for the immobilization of organelles and yeast. The chloroplasts immobilized by low temperature polymerization of various methacrylates and

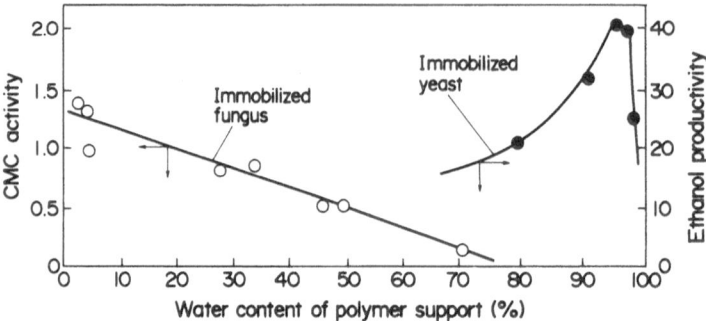

Fig. 14. Effect of hydrophilicity of the polymeric support on the activity of immobilized cells after cell culture

acrylates showed increased stability for long storage. For example, it retained its O_2 uptake activity even after 40 days storage at 4 °C, while intact chloroplasts lost their activity after only one day [67]. Yeast immobilized by the low temperature polymerization of hydrophilic methacrylate showed enhanced growth in the gel to form colonies of growing cells and showed much higher catalytic activity for the bioconversion of glucose to ethanol [68]. For example, the immobilized yeast produced more than 13 times the quantity of ethanol than free yeast. The activity of the immobilized yeast increased with the increase of hydrophilicity of the polymer. The most optimum hydrophilicity and softness of the gel matrix enhanced the cell growth of yeast most remarkably.

The second type of immobilization was carried out on a hollow fibrous support such as cloth, gauze and non-woven fibrous material reinforced by radiation polymerization of dispersed monomer. It was found that this type of immobilization was suitable for filamentous fungi such as *Trichoderma reesei* and *Sporotrichum cellulophirum* as cellulase-producing cells [69–72]. For example, the immobilized *Trichoderma* cell in a polymer reinforced gauze and non-woven cloth showed higher cellulase production in a continuous cell culture than the free cell. There was an optimum porosity and hollowness for the enhanced cell growth of fungi. It was observed that a fungus first attached itself to the fibrous support by winding the stretched arm around a micro-end on the surface of support and then spread into the hollow surfaces between the networked structure in the gauze. The relatively large hollowness was necessary for a sufficient supply of oxygen and nutrient and also for the volume increase of the growing cells during the enhanced cell growth. A suitable density of the networked structure in the fibrous support was also necessary for the enhanced growth.

The third type of immobilization was applied to the tissue cells, because a tissue cell can grow favorably after attachment and adhesion to a support surface. Various factors affecting the tissue cell growth such as hydrophilicity, electrostaticity and microporosity in a polymeric surface were studied. Kaetsu and coworkers studied the effects of various surfaces prepared by radiation techniques on the cell culture of glial cells and liver cells [73, 74]. A surface irradiated by an electron beam showed a change in the contact angle and wettability and promoted cell growth. Grafted copolymer surfaces with poly-methyl methacrylate showed quite different structures according to the solvents used such as acetone and dimethylformamide. The surface of methyl meth-acrylate-grafted PVF (polyvinylidene fluoride) in the presence of acetone had a more micro-heterogeneous surface and a more enhanced cell attachment and growth in comparison with a more homogeneous smooth surface obtained by the grafting in the presence of dimethylformamide. The cell growth had an increased rate at a certain optimum grafting ratio.

Thus, the surfaces prepared by radiation techniques such as irradiation, grafting, etching, and pore-forming polymerization at low temperatures provide useful materials for cell cultures with better wettability, micro-porous hetero-geneous structure and increased positive charges on the polymeric surfaces.

The immobilization of non-living cells has also been studied by Kaetsu and Kumakura [75]. The immobilized glucose-isomerase cell was investigated for a fructose conversion process by the radiation polymerization method. The immobilized glucose-isomerase cell showed yields as high as 80–90% and long duration of activity for repeated use. The various antigen cells and micro-organisms were also immobilized by radiation polymerization at low temper-atures and tested for their activities. For example, Fujimaki cells, Raji cells, and E. Coli cells were immobilized and used for the antigen-antibody reactions [76]. It was proved that those immobilized antigen cells can be reacted with the antibodies and used for the enzyme immuno-assay effectively.

6 References

1. Chapiro A (1967) J Polymer Sci C 16: 767
2. Chapiro A, Peree L (1967) Compt Rewnd C 264: 285
3. Kargin V, Kavanov VA, Papissov I (1967) J Polymer Sci C 4: 767
4. Hardy Gy, Nyitray K, Kevacs G, Federova N, Varga I (1962) Proceedings of Tihany Symposium on Radiation Chemistry, Sept 1962, Hungary, p 205
5. Barkalov IM, Goldanski VI, Enikolopov NS, Terekhova SF, Trofimova GM (1962) Dokl Acad Nauk SSSR 147: 395
6. Lando JB, Moravetz H (1964) J Polymer Sci C 4: 789
7. Kaetsu I, Tsuji K, Hayashi K, Okamura S (1967) J Polymer Sci A1 5: 1899
8. Kaetsu I, Kamiyama H, Hayashi K, Okamura S (1969) J Macromol Sci-Chem A3(8): 1509
9. Kaetsu I, Okubo H, Ito A, Hayashi K (1972) J Polymer Sci A-1 10: 2661
10. Kaetsu I, Ito A, Hayashi K, Okubo H (1972) J Polymer Sci A-1 10: 2203
11. Kaetsu I, Okubo H, Ito A, Hayashi K (1972) J Polymer Sci A-1 10: 2215
12. Kaetsu I, Ito A, Hayashi K (1973) J Polymer Sci Chem Ed 11: 1149
13. Kaetsu I, Ito A, Hayashi K (1973) J Polymer Sci Chem Ed 11: 1811
14. Kaetsu I, Yoshii F, Okubo H, Ito A (1975) Polymer Preprint 16: 465
15. Yoshii F, Okubo H, Kaetsu I (1978) J Applied Polymer Sci 22: 289
16. Kaetsu I, Kumakura M, Yoshii F, Yoshida K, Nishiyama S, Abe O, Tanaka H, Nakamura (1985) Radiat Phys Chem 25: 879

17. Dobo J (1970) Acta Chim Acad Hung 63: 435
18. Kawashima K, Uemeda K (1974) Biotechnol Bioeng 16: 609
19. Maeda H, Suzuki H, Yamauchi A (1973) Biotechnol Bioeng 15: 872
20. Kaetsu I, Kumakura M, Yoshida M (1979) Biotechnol Bioeng 21: 847
21. Kaetsu I, Kumakura M, Yoshida M (1979) Biotechnol Bioeng 21: 863
22. Kaetsu I, Kumakura M, Fujimura T, Yoshida M, Asano M, Kasai N, Tamada M (1986) Radiat Phys Chem 27: 245
23. Kaetsu I (1981) Radiat Phys Chem 18: 343
24. Kaetsu I (1985) Radiat Phys Chem 25: 517
25. Kumata M, Kaetsu I (1984) Acta Chim Hungaria 116: 345
26. Pekala W, Rosiak J, Rucinska-Rybus A, Barczak K, Galant S (1986) Radiat Phys Chem 27: 275
27. Boccu E, Carenza M, Lora S, Palma G, Veronese FM (1981) Biotechnol Bioeng 15: 1
28. Ha HF, Wang GH, Wu JL (1985) Progress report for IAEA Research Agreement: No. 4065/RB
29. Hoffman AS, Schmer G, Harris C, Kraft WG (1972) Trans Amer Soc Artif Int Organs 18: 10
30. Hoffman AS, Gomboltz WR, Uenoyama S (1986) Radiat Phys Chem 25: 375
31. Liddy JM, Garnett JL, Kenyon RS (1975) J Polymer Symp 49: 109
32. Beddows CG, Guthrie JT, Abdel-Hey FI (1981) Biotechnol Bioeng 23: 2885
33. Gao DY, Yoshida M, Kaetsu I (1988) Eur Polym J 24: 1037
34. Kaetsu I, Kumakura M, Asano M, Yamada A (1980) J Biomed Mater Res 14: 199
35. Kumakura M, Kaetsu I (1983) Appl Biochem Biotechnol 8: 87
36. Kumakura M, Kaetsu I (1984) Intern J Appl Radiat Isotop 35: 471
37. Kumakura M, Kaetsu I (1984) Anal Chim Acta 161: 109
38. Kumakura M, Kaetsu I (1984) Immunol Comm 13: 199
39. Catt K, Niall SD, Tregear GW (1967) J Lap Clin Med 70: 820
40. Rembaum A, Yen SPS, Cheong E, Wallance S, Molday RS, Gordon IL, Dreyer WJ (1976) Makromolecules 9: 328
41. Duval D, Collin C, Gaussens G, Nicaise M (1987) IAEA-TECDOC-486, p 129 Vienna
42. Leininger RI, Cooper CW, Falb RD, Geode GA (1966) Science 152: 1625
43. Kaetsu I, Yoshida M, Kumakura M, Yamada A, Sakurai Y (1980) Biomaterials 1: 17
44. Kaetsu I, Yoshida M, Yamada A (1980) J Biomed Mater Res 14: 185
45. Kaetsu I, Yoshida M, Asano M, Yamanaka H, Imai K, Mashimo T, Yuasa H, Suzuki K, Katakai R, Oya M (1988) J Controlled Release 6: 249
46. Asano M, Yoshida M, Kaetsu I, Oya M, Imai G, Mashimo T, Yuasa H, Yamanaka H, Suzuki K (1985) Polymer Preprints Japan 42: 783
47. Asano M, Yoshida M, Kaetsu I (1982) Polymer Preprints Japan 39: 327
48. Asano M, Yoshida M, Kaetsu I (1982) Polymer Preprints Japan 39: 612
49. Asano M, Yoshida M, Kaetsu I (1983) Polymer Preprints Japan 40: 525
50. Asano M, Yoshida M, Kaetsu I, Katakai R, Imai K, Mashimo T, Yuasa H, Yamanaka H, Suzuki K (1985) Biomaterials Japan 3: 85
51. Asano M, Yoshida M, Kaetsu I, Katakai R, Imai K, Mashimo T, Yuasa H, Yamanaka H, Suzuki K (1986) Biomaterials Japan 4: 65
52. Asano M, Yoshida M, Kaetsu I (1982) Polymer Preprints Japan 39: 333
53. Kaetsu I, Yoshida M, Asano M, Yamanaka H, Imai K, Yuasa H, Mashimo T, Suzuki K, Katakai R (1988) J Controlled Release 6: 249
54. Kaetsu I (1988) J Bioactive and Biocompatible Polymers 3: 164
55. Kaetsu I, Yoshida M, Asano M, Yamada A, Imai K, Nakai K, Mashimo T, Yuasa H (1988) Drug Develop Ind Pharm 14: 2519
56. Nakamura M, Hanu F, Yamada A, Sakurai Y, Yoshida M, Kaetsu I (1980) Cancer and Chemotherapy 7: 1824
57. Imai K, Yamanaka H, Yoshida M, Asano M, Kaetsu I, Yamazaki I, Suzuki K (1986) The Prostate 8: 325
58. Yamanaka H, Nakai K, Yuasa H, Imai K, Mashimo T, Kaetsu I, Yoshida M, Asano M (1986) Hormones and Clinic 34: 95
59. Yoshida M, Asano M, Kaetsu I, Imai K, Mashimo T, Yuasa H, Yamanaka H, Kawaharada U, Suzuki K (1987) Biomaterials 8: 124
60. Tanaka T, Nishio I, Sun ST, Ueno-Nishio S (1982) Science 218: 467
61. Kim SW, Okano T (1988) Proc Intern Symposium on Controlled Release of Bioactive Mater: Basel
62. Ishihara K, Kobayashi M, Ishihara N, Shinohara I (1984) Polymer J 16: 625
63. Hoffman AS, Arfassiabi A, Dong LC (1986) J Controlled Release 4: 213

64. Irie M (1990) Advances in Polymer Science 94: 28
65. Kaetsu I, Naka Y, Kodama J, Otori A, Morita Y (1989) On-off Switching Controlled Release System with Biosensor, 16th Intern Symposium on Controlled Release of Bioactive Materials, 6–9 Aug 1989, Chicago
66. Kaetsu I, Morita Y (1990) Intern J Artif Organs 85: 75
67. Fujimura T, Yoshii F, Kaetsu I (1980) Z. Naturforsch 35c: 477
68. Fujimura T, Kaetsu I (1981) Z. Naturforsch 37c: 102
69. Kumakura M, Kaetsu I (1984) Biotechnol Bioeng 26: 17
70. Kaetsu I, Kumakura M, Fujimura T, Kasai N, Tamada M (1987) Radiat Phys Chem 29: 191
71. Tamada M, Kasai N, Kumakura M, Kaetsu I (1986) Biotechnol Bioeng 28: 1227
72. Tamada M, Kasai N, Kaetsu I (1987) J Ferment Technol 65: 73
73. Yoshii F, Kaetsu I (1983) J Appl Biochem Biotechnol 8: 115
74. Yoshii F, Kaetsu I (1983) J Appl Biochem Biotechnol 8: 505
75. Kaetsu I, Kumakura M, Adachi S, Kikuchi S, Suzuki M (1983) Z Naturforsch 38c: 821
76. Kaetsu I, Kumakura M, Adachi S, Kikuchi S, Suzuki M (1983) Z Naturforsch 38c: 812

Editor: S. Okamura
Received December 11, 1991

Radiation Effects of Ion Beams on Polymers

Seiichi Tagawa
Research Center for Nuclear Science and Technology, University of Tokyo,
Tokai-mura, Ibaraki-ken 319-11, Japan

Recent progress in the radiation effects of ion beams on polymers are reviewed briefly. Our recent work on the radiation effects of ion beams on polystyrene thin films on silicon wafers and time resolved emission studies on polymers are described.

1 Introduction

Recently the number of papers about radiation effects of ion beams on polymers has been increasing very rapidly both in the fundamental and applied fields. A fairly large number of papers have been published on the fundamental aspects of radiation effects of ion beams on polymers, including high density electronic excitation effects [1, 2]. A number of papers have been published on the more applied aspects of the ion beam assisted advanced science and technology of polymers; examples of these are ion beam modification [1, 2] and lithography [3].

The present paper reviews the application of ion beams with energy above 10 keV to polymers and describes our recent work on the radiation effects of ion beams on solid polystyrene films studied by solubility change and time-resolved spectroscopy.

2 Application of Ion Beams to Polymers

2.1 General

The main application of ion beams with energy above 10 keV to polymers is reviewed here, although ion beams with energy below 10 keV are also widely applied in many fields such as ion beam etching and deposition.

Radiation effects on polymers have been studied for more than 30 years [4, 5, 6]. Most of the work on radiation effects on polymers has been carried out by using high energy photon (gamma-ray) and electron sources, since polymers are sensitive to any kind of ionizing radiation [7]. Even non-ionizing radiation such as ultraviolet and visible light excites electronic excited states in polymers and then photo-chemical reactions of polymers are induced from these electronic excited states. Studies on the radiation effects of other ionizing radiation on polymers have not been so popular for a long time. Recently, application of ion beams to polymers has been worthy of notice in fields of advanced science and technology, since ion beams induce different radiation effects, especially high density excitation phenomena [1, 2] induced by high energy gamma-rays and electrons. Research fields in the application of ion beams to polymers has now become very wide. They are, for example, fundamental research such as LET (Linear Energy Transfer) effects and changes in chemical bonds, mechanical properties, formation of nuclear tracks, carbonization or blackening, ablation, surface treatment, modification, deposition and etching.

2.2 LET Effects on Crosslinking, Main Chain Scission, Solubility Change and Mechanical Properties

LET effects on polymers are very fundamental data in radiation effects of ion beams on polymers (LET, dE/dx: linear energy transfer, also denoted as stopping power). LET effects on polymers are roughly divided into three groups: (A) radiation effects become more pronounced with increasing LET, (B) radiation effects become less pronounced with increasing LET, and (C) radiation effects are almost independent of LET. Roughly radiation-resistant polymers such as polystyrene are in group (A) and radiation-sensitive polymers such as PMMA are in group (B). Saturated hydrocarbon polymers such as polyethylene are in group (C).

The 100 eV-yield of crosslinking, G(X), of polystyrene is much higher at high LET than G(X) at low LET for gamma irradiation 0 [8, 9]. Similar results have been obtained by fast neutron irradiation [10]. Generally G(X) of polystyrene is reported to increase with increasing LET with the exception of one paper [11].

The 100 eV-yield of main chain scission, G(S), of PMMA decreases slightly with increasing LET [12].

The radiation effects on saturated hydrocarbon polymers such as polyethylene and ethylene-propylene co-polymers are almost independent of LET.

Recently more details of LET effects on polystyrene have been studied. Overlapping effects of the ion-track on G(X) have been studied [13]. The G(X) of polystyrene was measured over a wide range of kinetic energies, which included the Bragg peak of LET (stopping power) [9]. The values of G(X) differ for the same LET value at higher and lower ion beam energy points [9]. Similar results are observed for main chain scission, ablation and positive-negative inversion of PMMA.

2.3 Carbonization and Blackening

Carbonization is one of the most drastic phenomena of electronic excitation effects induced by ion beams [1, 2]. The mechanisms of carbonization and the chemical structure of the carbon-rich layer are still controversial problems. Many spectroscopic techniques have been applied to studies of carbonization such as RBS [14], Raman [14], IR [15], ESR [16], ESCA [17], electron microscope [18, 19] and electron energy loss spectroscopy in normal and reflection mode [20].

With carbonization, coloration of the polymers occurs [21]. Mechanisms of coloration or blackening of polymers induced by ion beams have been studied [22, 23] and two different models of blackening processes have been proposed: direct knock-on of atoms from polymer chains by nuclear collision [22] and high density electronic excitation effects by an electronic excitation process [23].

2.4 Ablation

Ion beam induced ablation is one of the most important electronic excitation effects [1, 2]. Ablation phenomena occur both thermally and photochemically in many kinds of materials including polymers and biological systems irradiated by both ion beams and high power laser pulses. The mechanisms of ablation of polymers induced by high density electronic excitation have not been made clear yet.

2.5 Nuclear Tracks

Nuclear tracks produced by ion beams have been applied to many fields such as track detectors for cosmic rays, nuclear track filters, membranes for separation processes, and single-pore membranes [24]. However, the mechanisms of formation of nuclear tracks have not been elucidated so far.

2.6 Modification of Ion Implantation

Polymer films change their properties during ion implantation [25–29]. Carbonization mentioned above is one of the ion beam induced modification techniques for polymer surfaces.

3 Basic Studies on Product Analysis and Ion Beam Pulse Radiolysis of Polymers

3.1 General

Recently radiation effects of ion beams on polymers such as changes in physical properties have been gaining attention in fields of advanced science and technology, especially from commercial interests. However, a fundamental understanding of these radiation effects has not yet been obtained because of the complexity of the reactions. There are a lot of problems concerning the nature of the microscopic damage and the radiation chemistry such as LET effects, carbonization or blackening, and ablation as described above. Recent studies have been carried out mostly for polyimide (mainly Kapton), polystyrene, polytetrafluoroethylene (Teflon), polyvinylidene fluoride, polymers for nuclear track detectors (mainly CR-39), and resist materials for microlithography (PMMA, etc.). Fundamental research has been concentrated on polyimide and polystyrene. There are a lot of data on gamma-ray and electron beam induced radiolysis of polystyrene and a few data on the radiolysis of polyimide, although there are a lot of technological papers on radiation effects on polyimide. Identification and reactivities of many kinds of reactive intermediates such as ions and excited states have been studied in detail for irradiated solid polystyrene but not for polyimide. Data on short-lived reactive intermediates in irradiated solid polymers are also available for polystyrene but not for polyimide.

The present chapter describes mainly the radiation effects of various ion beams on spin-coated polystyrene and PMMA films studied mainly by product analysis and by nanosecond ion beam pulse radiolysis.

Radiation effects of ion beams on polymers such as polystyrene have been studied using very quantitative, homogeneous, and energetically accurate irradiation data obtained by time-resolved and product analysis [30]. Recently main chain scission, ablative decomposition, and positive-negative inversion of PMMA induced by various ion beams have been investigated. The dependence of the beam energy and atomic number of incident ion beams on radiation effects has been considered.

Direct measurement of short-lived reactive intermediates by time-resolved spectroscopic methods is very important for understanding the detailed mechanisms of radiation effects. Very recently a new ion beam pulse radiolysis system using optical multi-channel detection has been developed. Although the use of ion beam pulse radiolysis for studying the radiation effects of ion beams on polymers was first reported by us [3, 30], the new system is highly modified for investigating ion beam reactions. Electron beam pulse radiolysis was also carried out complementarily.

In product analysis, G-values of the crosslinking were measured. The G-value is defined as the number of products or events per absorbed energy of 100 eV. Because charged particles lose kinetic energy at varying rates as they traverse matter, the "differential" G-value at a definite point (a definite kinetic energy) varies along the incident particle trajectory. Thus, in general, only the "averaged" G-value can be measured. In this experiment, however, by using polystyrene films thin enough so that the loss of the kinetic energy of the ion in traversing the film can be neglected, the differential G-values of crosslinking were measured easily and directly. The measurements were carried out for H^+, He^+ and N^+ ions (0.4–3.3 MeV) and also for 20 keV electrons. The process could be described in more detail with differential G-values than with averaged G-values.

3.2 Polystyrene

The investigation of the ion beam interaction with polystyrene by means of ion beam pulse radiolysis has the advantage that the reactive intermediates can be directly detected. Time profiles of the excimer fluroescence from ion irradiated polystyrene were measured using the polystyrene thin films. Thus, the transient phenomenon excited by the ion with a definite kinetic energy was observed.

3.2.1 Ion Irradiation Techniques

A monodispersed polystyrene sample with a weight-average molecular weight of about 9800 was used in this experiment. The polystyrene was spin-coated on a Si wafer and the thickness was 0.5 μm, as shown in Fig. 1a. The loss of kinetic energy of the ion in traversing the polystyrene thin film was estimated from the stopping power of polystyrene [31], and found to be almost negligible compared with the incident ion kinetic energy except for the case of low energy He$^+$ and N$^+$ ions. Thus, differential G-values of crosslinking could be measured in this experiment, as was mentioned in Sect. 1, while those for low energy He$^+$ and N$^+$ ions were somewhat averaged values.

Ion beams were obtained from a single-ended Van de Graaff generator (terminal voltage 0.4–3.75 MV) at the High Fluence Irradiation Facility, University of Tokyo.

The polystyrene films were irradiated with the ions in a vacuum chamber ($< 10^{-6}$ Torr) at room temperature. The sample holder had a hole 2 mm or 7 mm in diameter (Fig. 1b). The ion beam currents used in this experiment were very low (< 1 nA).

The film thickness before and after development were measured by Talystep (Taylor–Hobson). The normalized thickness was defined as the ratio of the thickness after irradiation to that before.

3.2.2 Solubility Change

Figure 2 shows the typical plots of the normalized thickness versus the irradiation dose in μC/cm^2 for polystyrene films irradiated with 1 MeV H$^+$, He$^+$, N$^+$ and 20 keV electron. The normalized thickness corresponds to the fraction of the insoluble part (gel fraction).

According to the statistical theory of crosslinking [33], when the initial molecular weight distribution is the random distribution, crosslinks are produced at random, and intramolecular linkage can be neglected, the behavior of gelation is expressed by the following equation (Charlesby-Pinner relationship),

$$s + s^{1/2} = 1/(x \times P_w), \quad s = 1 - g \tag{1}$$

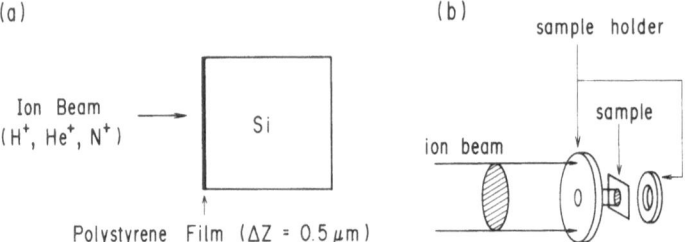

Fig. 1a, b. Schematic diagram of (a) sample and (b) sample holder. From Ref. 3

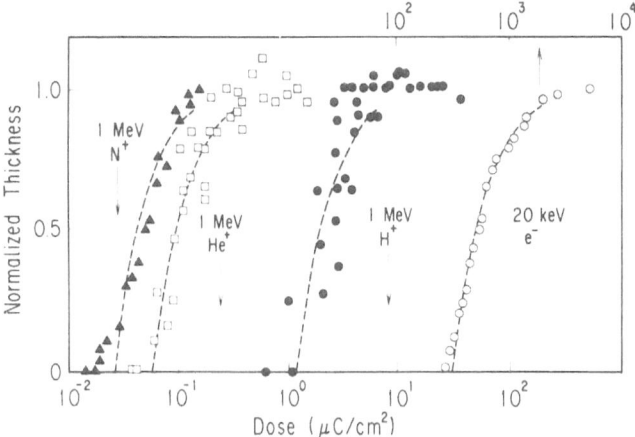

Fig. 2. Plots of the normalized thickness versus the irradiation dose in μC/cm² for polystyrene resist films irradiated with 1 MeV H⁺, He⁺, N⁺ and 20 keV electron. The *dashed lines* are the best fit curves for the Charlesby–Pinner relationship (Eq. (2) in the text). From Ref. 30

where g is the gel fraction, s the sol fraction, x the density of crosslinks, and P_w the weight-average degree of polymerization before irradiation. Since the density of crosslinks, x, is usually proportional to the irradiation dose, D, the following equation is obtained,

$$(1 - g) + (1 - g)^{1/2} = 2/(D/D_g^*) \tag{2}$$

where D_g^* is the gel point dose (the dose at g = 0 in Eq. (2)). From fitting Eq. (2) to the experimental data (Fig. 2), the gel point doses, D_g^*, were obtained. The gel point dose is related to the G-value of crosslinking as follows

$$G_x = 0.48 \times 10^6/(D_g^* M_w) \tag{3}$$

where G_x is the G-value of crosslinking (the number of crosslinks per 100 eV absorbed energy), D_g the gel point dose in Mrad, and M_w the weight-average molecular weight before irradiation. The differential G-values of crosslinking calculated from Eq. (3) are shown in Fig. 3 as a function of the stopping power. It should be noted that the G-values obtained in this experiment may contain some error in the measurement of the ion beam current. This was due to, for example, the instability of the ion beam current, while the beam sweeper was being used. The typical error is shown in Fig. 3.

As shown in Fig. 3, there is a large LET effect. The differential G-values for ions are larger than those for gamma-rays and fast electrons. This is considered as being due to the effect of the high density electronic excitations caused by ion beam impact. For He ions, the differential G-values are replotted against the kinetic energy in Fig. 4, as well as the stopping power. Each curve is normalized with the maximum value. The differential G-value curve has a sharp peak

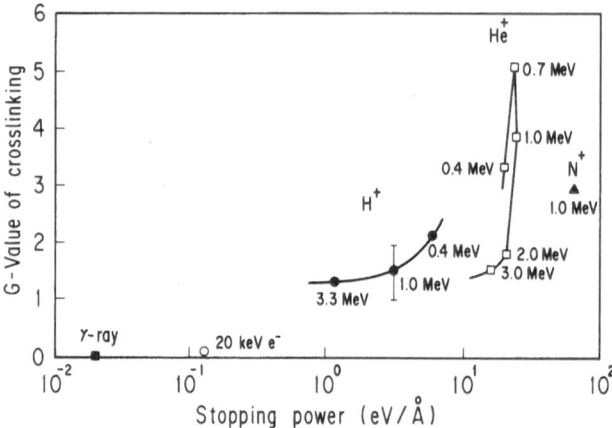

Fig. 3. The relation between differential G-values of crosslinking in polystyrene resist films and stopping powers. The G-values for gamma-ray irradiation (■) are from Ref. 34. The typical error is shown for 1.0 MeV H$^+$ beams. From Ref. 3

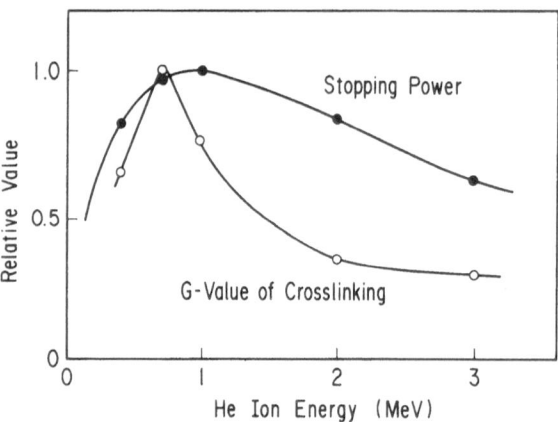

Fig. 4. The changes of the differential G-value of crosslinking in polystyrene resist films and of the stopping power in polystyrene resist films as a function of He ion energy. From Ref. 3

around 0.7 MeV and changes to a small extent in the high energy region, compared with the change in the low energy region. In this energy region, the shape of the differential G-value curve is very different from that of the stopping power curve. In addition, although He ions with energy of 2 MeV and 0.4 MeV have almost the same stopping powers (21.0 and 20.5 eV/A, respectively), the differential G-values for these ions are quite different. These results suggest that the effect of the high density electronic excitation is related not only to the stopping power but also to the structure of the track produced by the ion beam. To understand the effect of the high density electronic excitations better, ion beam pulse radiolysis studies have to be carried out. These are described in the next section.

3.2.3 Ion Beam Pulse Radiolysis [35, 36]

The pulse radiolysis method is a powerful means of studying the kinetics in radiation chemistry. We investigated the ion beam interaction with polystyrene using this method. It is a unique system, because pulse radiolysis is usually an electron pulse radiolysis.

The ion beam was obtained from single-ended Van de Graaff generator (see Sect. 2.1). This Van de Graaff generator is operated only in a DC mode. Thus, in order to carry out ion beam pulse radiolysis studies, an ion beam pulsing system is needed. In Fig. 5, the schematic diagram of the pulsing system is shown. By using both the chopper and the retracer, the DC ion beam from the Van de Graaff generator is converted to the pulsed ion beam with the variable repetition rate of $4/2^n$ MHz $(2 < n < 13$ in this system) and the definite beam positioned on the target.

In Fig. 5, the schematic diagram of the ion beam pulse radiolysis system with an optical emission spectroscope is also shown. The emission produced by the pulsed ion beam impact is detected through a monochromator by a fast photomultiplier tube (PMT) operated in a counting mode. The time profile of the emission is obtained by a coincident measurement between a photon and a

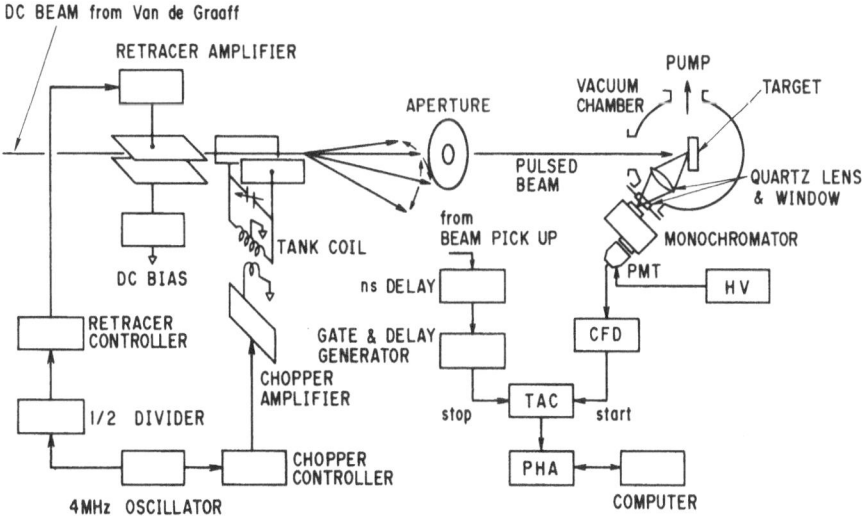

Fig. 5. The schematic diagram of the pulsing system and the ion beam pulse radiolysis system with an optical emission spectroscopy. *PMT* denotes photomultiplier tube; *HV*, high voltage supply; *CFD*, constant fraction discriminator; *TAC*, time to amplitude converter; and *PHA*, pulse height analyzer. From Ref. 36

pulsed beam, and finally accumulated in a pulse height analyzer (PHA) controlled by a personal computer. The output of the PMT is a start pulse of a time
to amplitude converter (TAC). As a stop pulse, the beam pick up signal is used.
In this system, three different pick ups are prepared; a capacitive pick up, a
secondary electron pick up and a retracer monitor, which is synchronized with
the output pluses of the retracer controller.

The time resolution of the system was measured and found to be about 1 ns.
This value was mainly dominated by the pulse width of the ion beam. In fact,
because in some cases the pulse width of the ion beam and the transit time jitter
of the ion have larger values caused by many factors (ion energy and species,
operating conditions of the accelerator and the pulsing system, the type of beam
pick up, etc.).

The same polystyrene thin films described in Sect. 2.1. were used. Thus, the
transient phenomenon excited by the ion with a definite kinetic energy was
observed.

3.2.4 Time Profiles of Polystyrene Excimer

In ion-beam irradiated polystyrene, some kinds of reactive intermediates are
produced. The excimer, which is one of the reactive intermediates, emits intense
fluorescence. Thus, we measured the time profiles of the excimer fluorescence
(328.5 nm) from ion irradiated polystyrene thin films. One of the results is shown
in Fig. 6a and b for irradiation with 0.6 MeV He^+ (several hundred pA beam
current). In Fig. 6a, the irradiation time dependence of the excimer fluorescence
intensity is shown. In Fig. 6b, the time profile (I) was recorded with an
irradiation time of 0 s–139 s (low dose-time profile), and (II) in the irradiation
time of 1839 s–3839 s (high dose-time profile). The following experimental
results were obtained.

(1) The intensity of the excimer fluorescence decreases with the irradiation
time, which suggests the existence of the quenching of the excimer by products.

(2) The decay of the low dose-time profile (I) is in agreement with that
obtained by the electron pulse radiolysis study of polystyrene [37, 38], the
lifetime of which was about 20 ns. The straight line in Fig. 6a corresponds to this
lifetime.

(3) On the other hand, the high dose-time profile (II) shows a different decay
from that by electron pulse radiolysis [37, 38].

The difference in (3) could be attributed to the quenching in (1). This quenching
process would take place at the region overlapped by two tracks formed by
separately incident ions, because the effects of the quenching process are found
to be much less in the low dose-time profile (I) than in the high dose-time profile
(II). Thus, this quenching process is referred to as intertrack quenching. There is
a possibility that the large LET effect mentioned in Sect. 2.2 is related to the
difference in the decay of the excimer fluorescence obtained by ion beam and

(a)

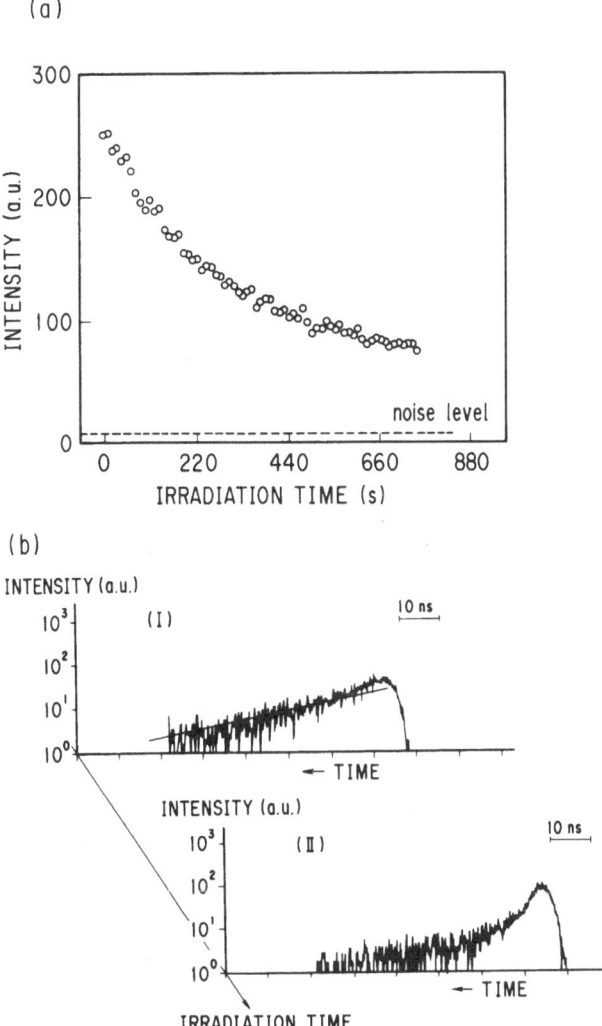

Fig. 6a. The irradiation time dependence of the intensity of the excimer fluorescence (328.5 nm) from a polystyrene thin film (0.5 µm thick) irradiated with 0.6 MeV He ions. From Ref. 36. (**b**) Time profiles of the excimer fluorescence (328.5 nm) from a polystyrene thin film (0.5 µm thick) irradiated with 0.6 MeV He ions. The low dose-time profile (I) was recorded in the irradiation time of 0 s–139 s, and the high dose-time profile (II) in the irradiation time of 1839 s – 3839 s. The *straight line* corresponds to the fluorescence lifetime obtained by the electron pulse radiolysis study [38]. From Ref. 36

electron pulse radiolysis, although there is also a possibility that the large LET effect is related to early events before the formation of the excimer. Unfortunately, the time resolution of our system is not sufficient to observe such early events.

The time profiles of the excimer fluorescence which were not influenced by the intertrack quenching just mentioned were also measured. Such time profiles were obtained under the low beam current condition that the intensity of the excimer fluorescence was not changed during the measurement.

The measured time profiles are shown in Fig. 7.

For the investigation of the LET effect on the time profile, it is important to measure the time profiles in the following three energy regions for the same kind of ions (see the stopping power curves in Fig. 8).

(1) The energy region higher than (2). High energy region.
(2) The energy region where the electronic stopping power has its maximum value. Medium energy region.
(3) The energy region lower than (2). Lower energy region.

In the theoretical treatment of the stopping power, the energy is generally divided into the three regions above. In order to measure the time profiles in these three energy regions for the same kind of ions, the kinetic energy of the ion

New ion accelerator facility

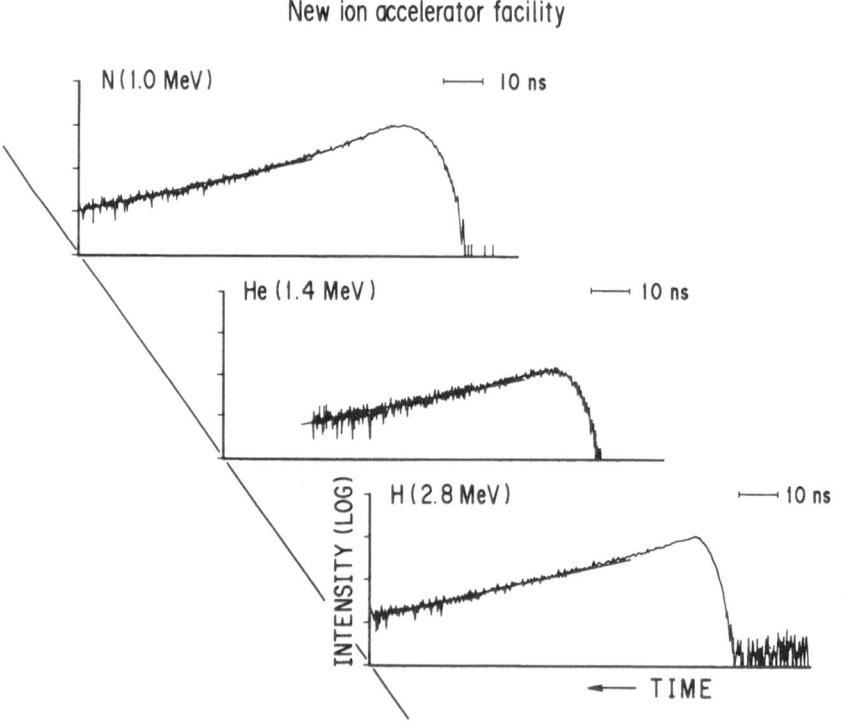

Fig. 7. Time profiles of the excimer from polystyrene resist films (0.5 μm thick) irradiated with ions. These time profiles were not influenced by the quenching seen in Fig. 6. The *straight lines* correspond to the fluorescence lifetime obtained by the electron pulse radiolysis study of polystyrene [38]. From Ref. 35

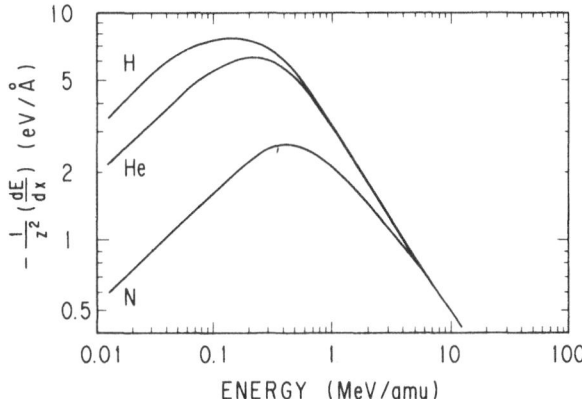

Fig. 8. The electronic stopping powers of polystyrene as a function of the ion kinetic energies. z is the atomic number of the projectile. The stopping powers were calculated from those of carbon and hydrogen (Northcliffe and Schilling, 1970) based on the additivity rule

should be varied over two orders. In our facility, this is impossible even if both the RF ion source and the PIG ion source are used. Thus, in this experiment, the ion species were changed as well as the energy. In Fig. 7, 2.8 MeV hydrogen (stopping power 1.4 eV/A) corresponds to the energy region (1), 1.4 MeV helium (stopping power 24 eV/A) corresponds to the energy region (2), and 1.0 MeV nitrogen (stopping power 68 eV/A) corresponds to the energy region (3). See the stopping power curves in Fig. 8.

Roughly speaking, the time profiles in Fig. 7 show single exponential decay, the lifetimes of which are in agreement with those obtained by the electron pulse radiolysis study of polystyrene [37, 38] (about 20 ns). The straight lines in Fig. 7 correspond to these lifetimes. These experimental results suggest that the excitation density of the excimer is not so high and the interaction between excimers is not strong enough to disturb the single exponential decay to any significant amount, in spite of the much higher stopping powers for ions than those for electrons. We could explain the above in terms of the quenching of the precursor of the excimer by nearby transient species produced through the high density electronic excitation in the core part of the track. This explanation is based on the theoretical treatment of the usual scintillators [39]. It should be noted that the quenching mentioned here is different from the intertrack quenching seen in Fig. 6.

Looking at decay curves in Fig. 7 closely, the small deviation from the single exponential decay can be observed. The fast decay component is observed in each time profile. The deviation is clear especially in the case of 1.0 MeV nitrogen. Because of the time profiles in Fig. 6, it is considered that the fast decay component is not due to the intertrack quenching but due to other effects of the high density electronic excitation. Further study of the fast decay component is in progress.

3.2.5 A New Emission Band Observed in Irradiated Polystyrene Films

Figure 9 shows the emission spectrum from a polystyrene film (0.5 μm thickness) excited by 2.5 MeV He$^+$ impact. The vertical axis was not calibrated for the efficiency of the optical system and the detector. The broad band at about 325 nm is the excimer fluorescence band. The intertrack quenching effect seen in Fig. 6 was not included in this emission spectrum, because the excimer fluorescence intensities were equal to each other within the fluctuation in two successively measured spectra. It should be noted that there was a weak band at about 500 nm. This new band was reported very briefly [35], although no assignment of this band has been given so far. This emission band is always observed for many kinds of irradiated polystyrene. The emission intensity of this band is very weak for fresh samples and increases with the irradiation at an early stage and then remains constant. It suggests that this band is due to radiolytic transient products of polystyrene with a long lifetime.

3.3 Polyethylene Model Compounds (n-Alkanes)

*Very recently, LET effects on fluorescence lifetimes of low molecular polyethylene model compounds (n-*alkane*) have been studied by many kinds of pulse radiolysis – methods such as electron beam, ion beam and synchrotron radiation (SR) [40] pulse radiolysis techniques [41]. Figure 10 shows time profiles of the fluorescence from neat n-*dodecane *liquids irradiated many kinds of radiation with different LET. The fluorescence lifetimes from irradiated neat*

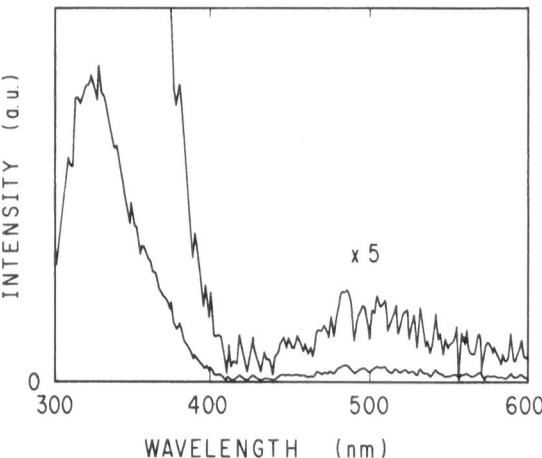

Fig. 9. The emission spectrum from a polystyrene resist film (0.5 μm thick) excited by 2.5 MeV He$^+$. The vertical axis was not calibrated for the efficiency of the optical system and the detector. This spectrum was not influenced by the quenching seen in Fig. 6

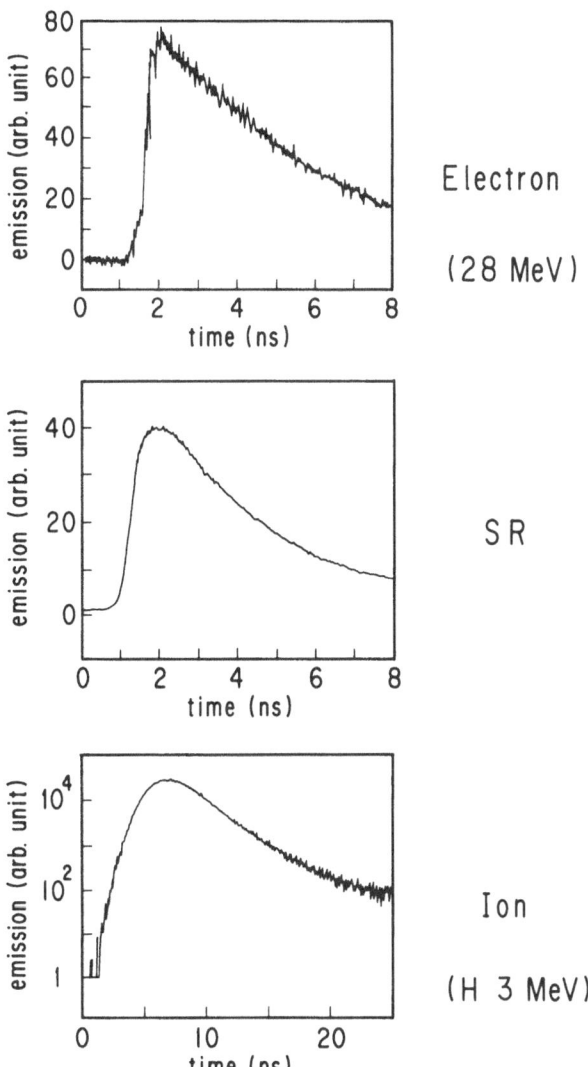

Fig. 10. Decay curves observed in electron beam, synchrotron radiation (SR), and ion beam pulse radiolysis of neat n-dodecane liquids

n-dodecane liquids decrease with increasing LET (Table 1). It is considered as being due to the interactions of reactive intermediates with each other.

3.4 PMMA

PMMA (polymethylmethacrylate) is a typical positive (scission type) electron beam resist for microlithography.

Table 1. Lifetimes of the fluorescence of neat *n*-dodecane liquids irradiated by 28 MeV electron beam, synchrotron radiation (SR) and ion beams

Irradiation sources	Lifetimes (ns)	LET eV/A
Electron (28 MeV)	4.3	0.02
SR (3–30 KeV)	3.6	0.1–1.0
Ion (3 MeV H)	3.2	0.7
Ion (1 MeV H)	2.8	2
Ion (3 MeV He)	2.3	9

3.4.1 High Density Excitation Effects

Recently spin-coated PMMA thin films with a thickness of 0.45 μm on silicon wafer were irradiated with various ion beams (H^+, He^+, N^+, Ni^{3+}). Ion beam energy regions are from 300 keV to 4 MeV. Irradiated PMMA films were developed by isopropyl alcohol in these experiments. After the irradiation by ion beams on PMMA in a vacuum, the thickness of the films were measured both before and after development. Various radiation effects on PMMA films such as ablation (sputtering), main chain scission, and positive-negative inversion were observed as shown in Fig. 11. These phenomena are very different from those in 60 Co gamma-ray or electron beam irradiation. Large LET effects are considered to be due to high density excitation by ion beams.

3.4.2 Ion Beam Pulse Radiolysis

Radiation effects of ion beams on PMMA films spin-coated on silicon wafers have been studied by ion beam pulse radiolysis. Figure 12 shows typical emission spectra of PMMA irradiated by 1 MeV nitrogen ion beams. The spectra changed drastically on irradiation.

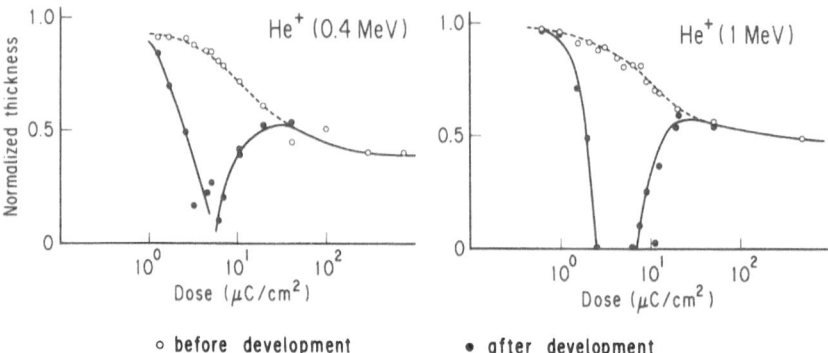

Fig. 11. The thickness of PMMA observed after 0.4 MeV He^+ and 1 MeV He^+ irradiation. Measurements of the thickness were before (○) and after (●) development

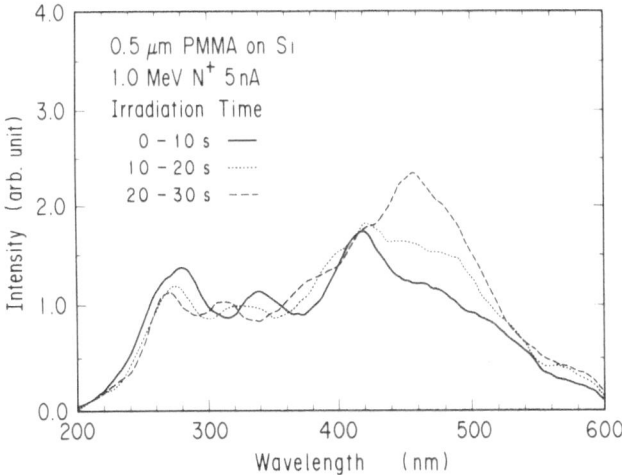

Fig. 12. Emission spectra from PMMA films on silicon wafers irradiated with 1 MeV nitrogen ions

4 Conclusion

The application of ion beams to polymers has been worthy of remark in the fields of advanced science and technology since the radiation effects of ion beams on polymers are different from those of conventional radiation such as electron beams and gamma-rays. The effects of ion beams are called LET effects but the detailed mechanisms of these effects on polymers have not been elucidated so far. So-called high density excitation effects such as carbonization, blackening, ablation and formation of nuclear tracks, which only occur at high densities, have been studied by a number of advanced spectroscopic methods.

Very recently LET effects of ion beams on both standard polymers such as polystyrene and low molecular polyethylene model compounds (*n*-alkanes) have been studied by time-resolved spectroscopic methods, that is, ion beam pulse radiolysis techniques. Further basic studies are necessary so that the detailed mechanisms of ion beams on polymers can be clarified, especially LET effects and high density excitation effects.

References

1. Venkatesan T (1985) Nucl Instr Meth Phys Res B7/8: 461
2. Brown WL (1986) Radiat Eff 90: 115
3. Selinger RL, Kubena RL, Olney RD, Ward JW, Wang V (1979) J Vac Sci Technol 16: 1610

4. Bopp CD, Sisman O (1951) ORNL: 1373; Bopp CD, Sisman O (1951) Nucleonics 13: 28
5. Charlesby A (1952) Proc Roy Soc A 215: 187
6. Dole M, Keeling CD, Rose DG (1954) J Am Chem Soc 76: 4304
7. Dole M (ed) (1972) The radiation chemistry of macromolecules, vols I, II, Academic, New York
8. Calcagno L, Foti G, Licciardello A, Puglisi O (1987) Appl Phys Lett 51: 21
9. Aoki Y, Kouchi N, Shibata H, Tagawa S, Tabata Y, Imamura S (1988) Nucl Instr Meth B33: 799
10. Egusa S, Ishigure K, Tabata Y (1980) Macromolecules, 13: 171
11. Schnabel W, Klaumunzer S (1989) Radiat Phys Chem 4: 323
12. Schnabel W, Klaumunzer S, Sotobayashi H, Asmussen F, Tabata Y (1984) Macromolecules 17: 2108
13. Licciardello A, Pugliti O, Calcagno L, Foti G (1989) Nucl Instr Meth B39: 769
14. Venkatesan T, Forrest SR, Kaplan MK, Schmidt PH, Murray CR, Brown WR, Wilkins BJ, Roberts Jr RF, Schonhorn H (1984) J Appl Phys 56: 2278
15. Puglisi O, Licciardello A, Pignataro S, Calcagno L, Foti G (1986) Radiat Eff 98: 161
16. Noda S, Hioki T (1984) Carbon 22: 359
17. Marletta G, Oliveri C, Ferla G, Pignataro S (1988) Surf Interf Anal 12: 447
18. Yoshida K, Iwaki M (1987) Nucl Instr and Meth B19/20: 878
19. Michael R, Stlik D (1987) Nucl Instr and Meth B28: 259
20. Marletta G, Pignataro S, Oliveri C (1989) Nucl Instr and Meth B39: 773
21. Venkatesan T, Calcagno L, Elman BS, Foti G (1987) In: Mazzoldi P, Arnold G (eds) Ion beam modification of insulators, Elsevier, Amsterdam, p 301
22. Fink D, Muller M, Chadderton LT, Cannington LH, Elliman RG, McDonald DC (1988) Nucl Instr and Meth B32: 125
23. Davenas J, Boiteux G, Fallavier M (1989) Nucl Instr and Meth B39: 796
24. Eberhard B, Spohr R (1983) Rev Modern Phys 55: 907
25. Mazurek H, Day DR, Maby EW, Abel JS, Senturia SD, Dresselhaus MS, Dresselhaus G (1983) J Polm Sci Polym Phys Ed 21: 537
26. Forrest SR, Kaplan ML, Schmidt PH, Venkatesan T, Lovinger AJ (1982) Appl Phys Lett 41: 708
27. Yoshida K, Iwaki M (1987) Nucl Instr and Meth B19/20: 878
28. Hoiki T, Noda S, Sugiura M, Kakeno M, Yamada K, Kawamoto J (1983) Appl Phys Lett 43: 30
29. Wada T, Takeno A, Iwaki M, Sasabe H (1987) Synthetic Metals, 18: 585
30. Tagawa S, Kouchi N, Shibata H, Tabata Y (1989) Adv in Resist Technology and Processing VI, SPIE Symp 1086: 65
31. The stopping powers of polystyrene for ions were calculated from those of C and H [32a] based on the additivity rule. For 20 KeV electrons, the stopping power was calculated from that for 10keV electrons [32b] based on the Bethe-Bloch formula.
32. (a) Northcliffe LC, Schilling RF (1970) Nucl Data Table, A7 233 (b) Ashley JC, Tung CJ, Ritchie RH (1978) IEEE Trans Nucl Sci NS-25, 1566
33. Saito O (1973) In: Dole M (ed) The radiation chemistry of macromolecules, Vol 1, Academic, New York, p 238
34. Parkinson WW, Keyser RM (1973) In: Dole M (ed) The radiation chemistry of macromolecules, Vol 2, Academic, New York, p 72
35. Kouchi N, Tagawa S, Kobayashi H, Tabata Y (1989) Radiat Phys Chem 34: 453
36. Kouchi N, Aoki Y, Shibata H, Tagawa S, Kobayashi H, Tabata Y (1989) Radiat Phys Chem 34: 759
37. (a) Tagawa S (1986) Radiat Phys Chem 27 455; (b) Tagawa S (1990) In: Tabata S, Mita I, Nonogaki S, Horie K, Tagawa S (eds) Polymers for microelectronics, Kodansha, VCH Weinheim, p 63
38. Itagaki H, Horie K, Mita I, Washio M, Tagawa S, Tabata Y (1983) J Chem Ohys 79: 3996
39. Voltz R, Lopes da Silva J, Laustriat G, Coche A (1966) J Chem Phys 45: 3306
40. Ogata A, Tagawa S (1989) Review Sci Instr 60: 2197
41. Yoshida Y, Shibata H, Tagawa S, Washio M, Tabata Y, Kouchi N, Ogata A (1989) Adv in Resist Technology and Processing VI, SPIE Symp 1086: 274
42. Shibata H, Tagawa S, Yoshida Y (1989) Preprint of the fall Symposium of the Polymer Society in Japan.

Polymer Materials for Fusion Reactors

H. Yamaoka
Research Reactor Institute, Kyoto University, Kumatori, Osaka 590-04, Japan

The radiation-resistant polymer materials have recently drawn much attention from the viewpoint of components for fusion reactors. These are mainly applied to electrical insulators, thermal insulators and structural supports of superconducting magnets in fusion reactors. The polymer materials used for these purposes are required to withstand the synergetic effects of high mechanical loads, cryogenic temperatures and intense nuclear radiation. The objective of this review is to summarize the anticipated performance of candidate materials including polymer composites for fusion magnets. The cryogenic properties and the radiation effects of polymer materials are separately reviewed, because there is only limited investigation on the above-mentioned synergetic effects. Additional information on advanced polymer materials for fusion reactors is also introduced with emphasis on recent developments.

1 Introduction

Design studies for several fusion reactors and devices have become more detailed in recent years. One of the important aims of design engineering is to select suitable materials which perform reliably under the most severe conditions [1].

Among several conceptual designs of fusion reactors, the machines based on magnetic confinement, such as Tokamak- and mirror-type reactors, employ a variety of superconducting magnets which are operated at cryogenic temperatures. In a D–T cycle, the fusion energy is liberated as kinetic energies of 3.5 MeV alpha particles and 14.1 MeV neutrons. The components of superconducting magnets are exposed to intense radiation of fast neutrons and secondary γ-rays. The radiation level at the position of magnets in fusion reactors mainly depends on the design of reactors, particularly on the composition and thickness of the blanket and the magnet shield [2]. Recent studies on radiation effects in fusion-magnet materials have shown that the maximum accumulated radiation levels to the magnets are of the order of 10^8 Gy at the end of plant life of 20 or 30 years [2, 3].

The greater tolerance for radiation damage in inorganic materials makes them attractive for fusion magnets, but their brittleness and the difficulty of fabrication techniques presents a serious limitation to their practical uses. Therefore, the application of organic materials has been considered for electrical insulators, thermal insulators and a part of structural supports in the magnets [4–9].

The main requirements for electrical insulators used in superconducting fusion magnets are:

1. Excellent mechanical properties at cryogenic temperatures.
2. High radiation tolerance at cryogenic temperatures.
3. Good electrical properties at cryogenic temperatures.
4. Low level of volatile radiolysis products.
5. Low level of activated products by neutron irradiation.

At present it is very difficult to estimate the synergetic effects of high mechanical loads, cryogenic temperatures and high fluences of nuclear radiation on polymer materials. In this review, therefore, the effects of each above-mentioned factor on the polymer properties will be separately introduced.

2 Mechanical Properties at Cryogenic Temperatures

2.1 Conventional Polymers and Engineering Plastics

The available data on the mechanical properties of polymer materials at cryogenic temperatures have been reported for the last few decades mainly in close connection with space technology.

In general, most polymers lose their ductile properties below the glass transition temperatures (T_g), the point at which the movements of polymer chain segments become extremely restricted. In amorphous polymers, the characteristics of the low temperature relaxations are directly related to the chemical structure and the dynamics of polymer chains. There are several possible types of local molecular motions, such as torsional oscillations of polymer segments, crankshaft and kink motions accompanied by coordinated movements of a few chain bonds, and rotations of side chains or terminal groups. It has been also proposed that the local intermolecular rearrangements result in the relaxations of low temperature region. In crystalline and oriented polymers, the relaxation phenomena seem to be more complex, indicating considerable dependence on the morphology of polymers.

The early review by Woodward [10] has provided the data on mechanical loss measurements of several polyolefins and vinyl polymers at cryogenic temperatures. Although the mechanical strengths, such as tensile strength, compressive strength and Young's modulus, of most polymers increase or remain constant as the temperature is decreased, the elongation to failure decreases to extremely low values at cryogenic temperatures [11]. This behavior tends to limit the use of most polymers at cryogenic temperatures, particularly where the flexibility of polymers is required as in the case with wire and cable insulations.

Polyethylene terephthalate (PET) is well known to be one of typical polymers usable in cryogenic environments. The mechanical and relaxation behavior of PET at cryogenic temperatures has been extensively studied with dynamic mechanical [12] and stress-strain measurements [13]. There is general agreement on the existence of a pronounced relaxation maximum at around 200 K below T_g of PET. This broad and asymmetric peak is believed to be due to multiple relaxations attributed to motions of methylene groups and carboxylene groups in amorphous regions. Furthermore, the torsion pendulum experiments of crystalline and oriented PET specimens at cryogenic temperatures showed the presence of relaxation maxima at 46 and 26 K, respectively, which do not exist in the amorphous sample [12]. Tensile tests with biaxially oriented and heat-set films of PET at cryogenic temperatures revealed a twofold increase in elastic modulus and a tenfold increase in toughness over those of the amorphous film [13].

An extensive compilation and evaluation of mechanical, electrical, and thermal properties of six commercially available polymers was performed by Reed et al. [14]. It was shown in their summarized data that polypyromellitimide (PPMI), which is obtained by the polycondensation between pyromellitic acid and aromatic diamine, exhibits excellent mechanical properties at both high and low temperatures and retains ductility even at cryogenic temperatures, as seen in Fig. 1.

Earlier investigations on the dynamic mechanical properties of PPMI over a wide range of temperatures indicated the existence of two distinct relaxations at around 250 and 400 K [15, 16]. The former relaxation was assigned to adsorbed water molecules in the polyimide chain and the latter was due to local relaxation modes of the backbone. Recently Ahlborn reported the mechanical relaxation of various polymers at low temperatures [17]. In the film sample of PPMI, the small relaxation at 93 K attributed to the motion of phenyl rings was observed, although the dominant damping peak at 198 K is not yet explained.

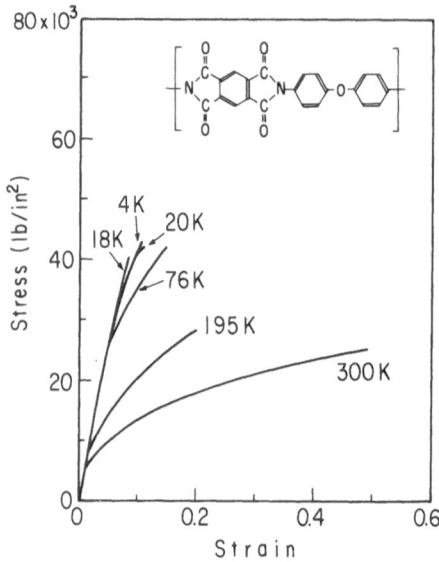

Fig. 1. Stress-strain curves of PPMI at different temperatures. (1 lb/in² = 6.895 kPa)

Polyphenylene sulfide (PPS), which is made from p-dichlorobenzene and sodium sulfide, is one of commercially available engineering plastics. The tensile experiments of PPS film at cryogenic temperatures were first carried out by Yamaoka and Miyata [18]. Figure 2 shows the stress-strain curves of PPS film at different temperatures. The curve at 300 K is typical of hard ductile polymers and shows a gradual increase in strength beyond the elastic limit. This feature remains unchanged with decreasing temperature, indicating that no transition region from ductile to brittle exists between 300 K and 4.2 K. In addition, the mechanical properties of PPS film measured at 77 K were found to be almost

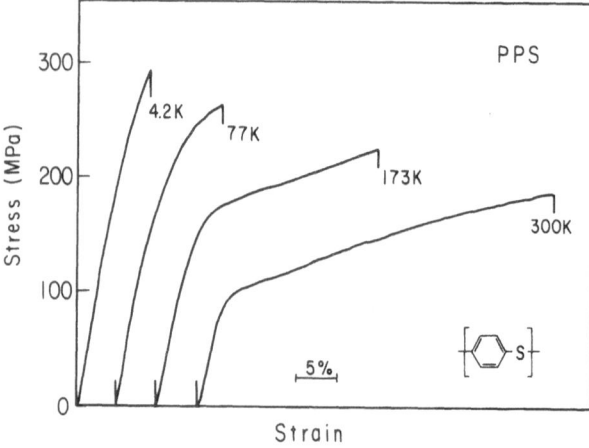

Fig. 2. Stress-strain curves of PPS at different temperatures

independent of the large variation in strain rate. These results indicate that PPS film possesses excellent mechanical properties at cryogenic temperatures.

2.2 Polymer Composites

Fiber reinforced plastics (FRP) are widely utilized now for a number of cryogenic applications. The matrices used are generally thermosetting resins such as epoxies, polyesters and phenolics, since they are easy to apply in impregnating processes.

The results of cryogenic experiments on a number of glass-fiber reinforced plastics (GFRP) in the early stage are summarized in the reports written by Chamberlain [19] and Hertz [20]. Figure 3 indicates the average tensile strength and modulus values of various laminates fabricated with glass cloth at room temperature and 20 K [19]. All matrices tested showed the increase in strengths and moduli at temperatures down to 77 K, with little or no further increase on cooling to 20 K. As seen from Fig. 3, the epoxy resins had the highest strengths and moduli, and a similar superiority was also observed in compressive, flexural, and bearing strengths.

The cooperative program in the United States was undertaken by the National Bureau of Standards, the National Electrical Manufacturer's Association (NEMA) and US laminating industry in order to establish specifications for cryogenic grades of glass-cloth-reinforced epoxy laminates [21]. The resulting materials, designated G-10CR and G-11CR, are high quality variants of NEMA G-10 and G-11 products to meet cryogenic performance. G-10CR is a

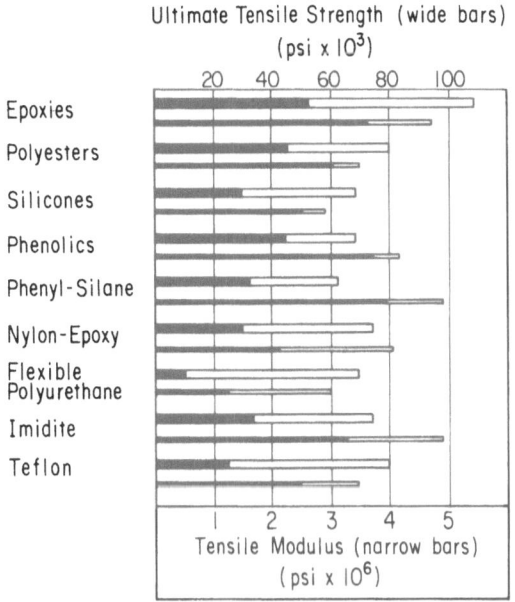

Fig. 3. Average ultimate tensile strength and modulus of various laminates with glass cloth at room temperature and at 20 K. *Shaded bars* indicate the values at room temperature (1 psi = 6.895 kPa)

heat-activated, amine-catalyzed, epichlorohydrin-bisphenol A solid-type epoxy resin laminate reinforced with continuous-filament E-glass having a silane finish. G-11CR is an aromatic-amine hardened, diglycidyl ether of bisphenol A liquid-type epoxy resin laminate reinforced in the same manner. Resin weight fraction is 32 to 36% for G-10CR and 28 to 33% for G-11CR. Several mechanical tests for both laminates were performed in the temperature range between 4 K and room temperature. The tensile strengths of G-10CR and G-11CR as a function of temperature are shown in Fig. 4, where the shaded areas indicate the extent of upper and lower data spread among tests [21]. G-11CR was found to show a higher tensile strength in the fill direction and a higher modulus in both warp and fill directions than those of G-10CR. Thus the CR-grade glass-epoxy laminates proved useful in applications that require commercially available structural or insulating materials of predictable performance at cryogenic temperatures.

From the viewpoint of cable insulation in superconducting magnets at cryogenic temperatures, it would be preferred that the insulating materials have similar coefficients of thermal expansion to those of superconductors. Since FRP are not homogeneous materials, their expansion properties are a combination of the individual properties of the filler and the resin. Hamelin first succeeded in developing high-density GFRP which exhibits a linear thermal contraction equal to that of superconductors with the use of epoxy resin preimpregnated glass-cloth laminates [22]. Fukushi et al. also prepared high-density GFRP by using a cross-piled unidirectional woven cloth with epoxy or bismaleimide-triazine (BT) resin [23]. The thermal contraction of the GFRP was nearly as low as that of stainless steel and the compressive strength at 77 K measured perpendicular direction to the lamination was found to be over 1,300 MPa.

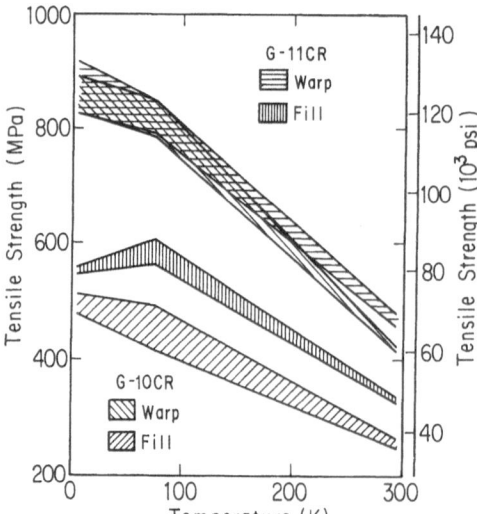

Fig. 4. Temperature dependence of tensile strength for G-10CR and G-11CR. *Error bands* show upper and lower data spreads. (1 psi = 6.895 kPa)

The conventional GFRP, which are reinforced in one or two directions with glass fibers, show the anisotropy of mechanical properties and thermal contraction in the direction perpendicular to the fibers, especially at cryogenic temperatures. Their low interlaminar shear strengths also result in the restricted applications at cryogenic uses. In order to improve the problems induced by the anisotropic properties, three-dimensional glass-fabric reinforced plastics (3DFRP) were recently developed by Nishijima and his coworkers [24–28]. Epoxy and BT resins were used as the matrices. These 3DFRP were found to show well-balanced mechanical properties such as Young's modulus, tensile, flexural, and compressive strengths and also show lower thermal contraction than the conventional GFRP in the thickness direction at cryogenic temperatures.

FRP made with similar techniques from graphite, boron and high-strength organic fibers also show the excellent performance in cryogenic environments. The detailed reviews by Kasen [29–32] and Hartwig [33–35] on the cryogenic properties of all types of FRP are worth studying. Furthermore, the recent topics on the cryogenic properties of various FRP were extensively introduced in the new volume of Advances in Cryogenic Engineering (Materials) [36] and the special issue of Cryogenics [37].

The summarized results from a number of reports revealed that even within the same combination of resins with fillers there is a large scattering in test data on cryogenic properties of FRP. This may be attributed to large number of factors such as resin content, relative flexibility of the resin system, curing temperature and pressure, cure and post-cure cycle, surface treatment of fillers, specimen configuration, as well as the apparatus and the conditions of testing. A great deal of work on the standardization in cryogenic properties of FRP still remains to be done.

3 Radiation Effects at Cryogenic Temperatures

A number of investigations on radiation effects of polymers at ambient temperature have been carried out and summarized in some recent publications [38–40]. However, there are very few experimental results on cryogenic temperature irradiation of conventional polymers. Figure 5 shows the summary of early experiments on irradiation of polymers at 77 K up to γ-ray dose of 2×10^6 Gy [41, 42]. As is evident from Fig. 5, the candidate polymers that can be used at cryogenic temperatures are only aromatic based epoxy resins, polyimides, and polystyrene. This means that the choice of polymers for superconducting magnets to be operated in a radiation environment is rather limited. Therefore, the intention of this chapter is to give information on the radiation tolerance of recently developed polymers.

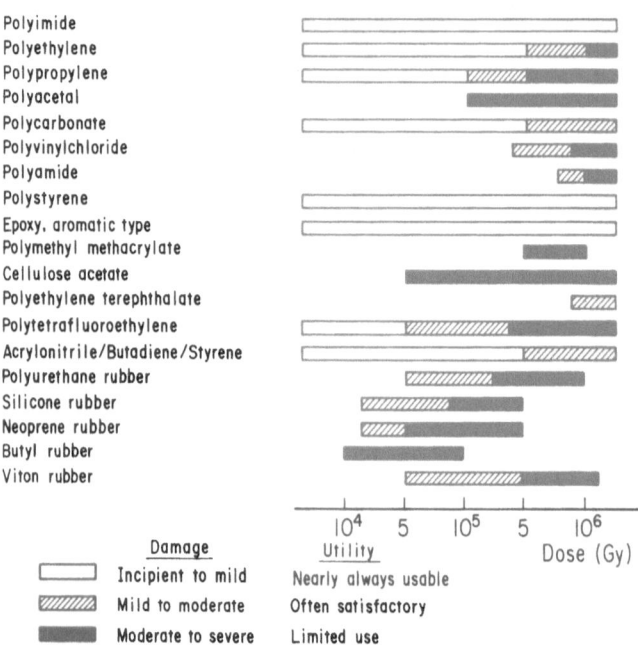

Fig. 5. Relative radiation resistance of various polymers at 77 K

3.1 Epoxy Resins

Because of their excellent processing characteristics, especially for impregnating wound coils or fabrics, epoxy resins have been widely used for cryogenic temperature applications. There are many different types of epoxy resins available.

Evans et al. [43] carried out 4 MeV electron irradiations of 14 different epoxy resins at 77 K which were selected from a large number of resin systems after screening tests on thermal shock at cryogenic temperatures [44]. The results of flexural tests show that most of these irradiated resins possess only moderate resistance to radiation. Takamura and Kato [45] tried to irradiate the bisphenol-A type epoxy resins with various hardeners at 5 K in a fission reactor and reported that the compressive strength of these epoxy resins decreased sharply after a combined neutron and γ-ray irradiation equivalent to a dose of about 10^7 Gy.

Among several kinds of mechanical properties, fatigue resistance is probably one of the most important design parameters for the superconducting magnet under pulsed mechanical loads. Figure 6 shows the effect of reactor irradiation at about 20 K on the cryogenic fatigue resistance of epoxy resin reported by Nishijima et al. [46]. The results of the fatigue tests indicate that the load decreases remarkably and the scatter of fatigue load increases after reactor

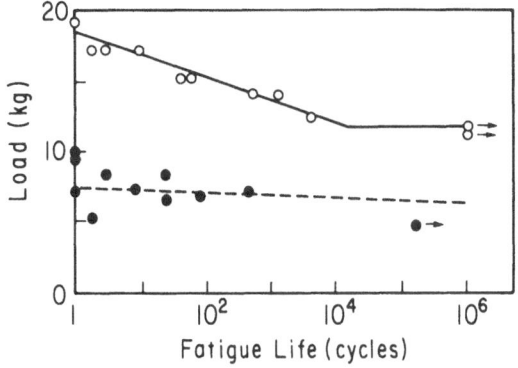

Fig. 6. Load-endurance diagrams of epoxy resin before and after reactor irradiation at 20 K. ($\bigcirc$) control; ($\bullet$) irradiation up to a fast neutron flux of 2.5×10^{20} n/m^2 and a gamma ray dose of 2.8×10^6 Gy

irradiation. These may be attributed to the decrease in average molecular weight of epoxy resin induced by irradiation.

Another radiation effect on organic insulators is the evolution of gaseous products by the degradation of polymers. These products are trapped in the polymers at cryogenic temperatures and released on warming of the specimens to room temperature after irradiation. It seems probable during the periodic warming of superconducting magnets that the sudden release of trapped gases causes the formation of microcracks or foam-like structures which would seriously reduce the mechanical properties of materials. Table 1 represents the results of gas analyses on irradiated epoxy resins reported by Morgan et al. [47]. Remarkable differences in the amount and the composition of gas evolved were observed in the resins cured with various hardeners. This result indicates that the formation of volatile products during irradiation is dependent upon not only on the kinds of resins but also on several manufacturing processes of resins.

Based on different data from the available literatures on irradiated epoxy resins at low temperatures, Evans and Morgan [48] proposed the following generalizations:

1. Acid anhydride hardeners, in general, result in brittle systems after curing with a relatively high level of gas evolution on irradiation.
2. Aliphatic amine hardeners make it possible to formulate 'tough' epoxide systems, but with limited resistance to high energy radiation.

Table 1. Gas evolution from DGEBA* epoxide resin cured with various hardeners

	Gas evolved (cm^3/g/Gy)	Composition (% volume)				
		H$_2$	CO$_2$	N$_2$/CO	CH$_4$	C$_2$H$_6$
Aromatic amine	2.5×10^{-7}	77.1	0.8	20.9	0.9	0.3
Aliphatic amine	4.3×10^{-7}	88.4	–	10.0	1.3	0.3
Acid anhydride	7.3×10^{-7}	19.9	56.9	23.2	–	–

*DGEBA: Diglycidyl ether of bisphenol A

3. Aromatic amine hardeners confer the greatest radiation stability, on a mechanical basis, and also release the lowest levels of gaseous degradation products.

3.2 Aromatic Polyimides

Aromatic polyimides are relatively new polymer materials, having been developed primarily for high temperature uses. PPMI, which is one of the typical aromatic polyimides, is commercially available as film called Kapton or H-film and as bulk material called Vespel. In qualitative experiments, Kapton film was found to be useful for cryogenic system at least up to a dose of 1×10^8 Gy [49]. After irradiation to 1×10^8 Gy at 4.9 K, Vespel loses only 8% of its initial flexural strength and shows a slight increase of 8% in compressive strength, measured at 77 K [50]. In Table 2, the results of gas analyses at two successive intervals evolved at 307 K after irradiation to 1×10^8 Gy at 4.9 K are given [50]. It is noted that the composition of off-gas changes remarkably with time after irradiation. A larger portion of the gas evolved after the first 24 hours is hydrogen, and the composition of gas evolved at later times has a higher abundance of heavier molecules indicating slower diffusion out of the material.

Recently Yamaoka and Miyata carried out the reactor irradiation at 20 K on two kinds of aromatic polyimides, Upilex-S and Upilex-R [51]. The stress-strain curves of irradiated Upilex-S films are depicted in Fig. 7. No essential changes in both elastic modulus and yield strength were observed for Upilex-S after irradiation up to 8×10^6 Gy, although a slight decrease of the ultimate elongation was detected. A similar tendency was found in the reactor irradiation of Upilex-R at 20 K, in spite of the fact that its structural unit of the main chain is different from that of Upilex-S. These results indicate that a homologue of aromatic polyimides exhibits the excellent radiation stability even at cryogenic temperatures as well as at ambient temperature.

Table 2. Gas evolution from polyimide at 307 K after irradiation up to 1×10^8 Gy at 4.9 K (10^{-4} grams gas/gram resin)

	Interval after irradiation (days)	
	0–1	1–6
H_2	14.3	0.2
CH_4	0.3	0.5
H_2O	0.6	7.4
N_2 and CO	16.0	23.4
O_2	0.02	0.1
CO_2	0.9	3.2
C_2H_4	0.2	0.4
Total	32.3	35.2

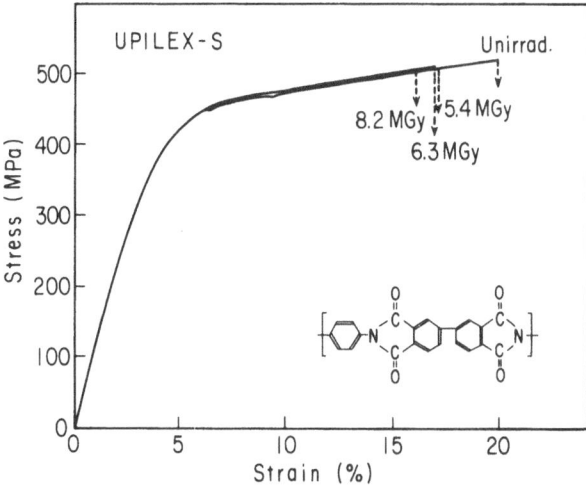

Fig. 7. Stress-strain curves of irradiated Upilex-S at 20 K

3.3 Other Aromatic Polymers

Polyethylene terephthalate film (Mylar) is often used at cryogenic temperatures for electrical- or thermal insulation, as described in the preceding section. However, the radiation tolerance of Mylar is rather poor as shown in Fig. 5. Takamura and Kato reported that Mylar was too brittle to handle after irradiation of 6.2×10^6 Gy at 5 K [45].

Figure 8 shows the effects of reactor irradiation at 20 K on the mechanical properties of polyphenylene sulfide (PPS) and polyethylene terephthalate (PET) films [52]. In the irradiated PET films, the tensile strength remarkably decreased with an increase in the absorbed dose above 2×10^6 Gy and the ultimate elongation gradually reduced with an increase of the dose, in accordance with the earlier data. On the other hand, both the tensile strength and the ultimate elongation of PPS films were substantially independent of irradiation dose up to 8×10^6 Gy. The tensile modulus of PPS film was found to be about 4.8 GPa and this value was also constant during the irradiation up to 8×10^6 Gy. These results indicate that PPS is one of the candidate materials as an insulator for superconducting magnets in fusion reactors.

Aromatic polyamide film (Nomex) was found to decrease slightly in tensile strength after irradiation of 6.2×10^6 Gy at 5 K [45], but confirmed still available for cryogenic uses at 1×10^8 Gy [49].

Bell and Pezdirtz reported that polyethylene-2,6-naphthalenedicarboxylate (PEN) exhibited extremely resistant properties to radiative degradation [53]. The tensile strength and the ultimate elongation of PEN film were retained in spite of exposure to γ-ray doses in excess of 1×10^7 Gy at ambient temperature.

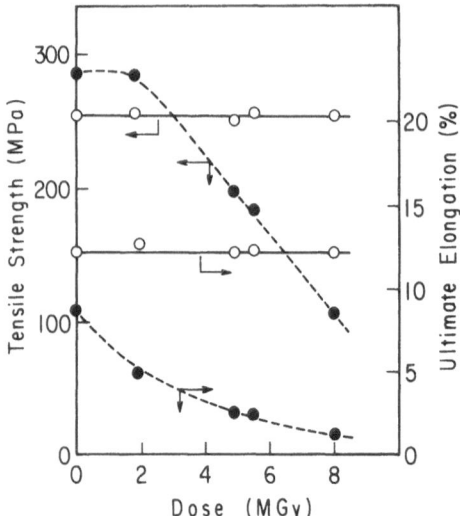

Fig. 8. Effect of reactor irradiation at 20 K on the mechanical properties of PPS and PET. (○) PPS; (●) PET

Havens and Bell performed an electron beam irradiation of methylene-bridged aromatic polyesters, which were synthesized by polycondensation between two pairs of aromatic diacid chlorides and 3,3'-methylenediphenol or 4,4'-methylenediphenol [54]. They found that irradiation of thin films of these polyesters at room temperature resulted in some chain extension and cross-linking, and that irradiation at a temperature near or above the glass transition temperatures of the polymers greatly enhanced the tendency for the polymers to crosslink.

In recent years, remarkable progress has been made in the syntheses of aromatic and heterocyclic polymers to search a new type of radiation resistant polymers. Sasuga and his coworkers extensively investigated the radiation deterioration of various aromatic polymers at ambient temperature [55–57] and reported the order of radiation resistivity evaluated from the changes in tensile properties as follows: polyimide > polyether ether ketone > poly-amide > polyetherimide > polyarylate > polysulfone.

The preliminary electron beam experiments on all aromatic poly(arylene ether sulfone)s were performed by Hedrick et al. [58] and the detailed investigation on radiation effects of these polymers were recently carried out by Lewis et al. [59]. They found that an aromatic polysulfone based on 4,4'-biphenol and 4,4'-dichlorodiphenyl sulfone exhibited the highest radiation tolerance in a systematic series of poly(arylene ether sulfone)s with γ-irradiation up to 4×10^6 Gy.

Recently Hegazy et al. investigated the gas evolution from various aromatic polymers by γ- and electron-irradiation under vacuum at ambient temperature [60]. Table 3 shows the summarized results of gas analyses evolved by γ-irradiation. The structure of the polymers evidently exerts great influence on the yield and the component of gases evolved. Furthermore, the gas evolution is

Table 3. Yield of evolved gases from aromatic polymers by gamma irradiation under vacuum

Component	Yield (10^{-7} mol/g·MGy)							
	Kapton	Uplx-R	PEEK-c	PEEK-a	PES	Uplx-S	U-PS	U-Polym
Total	1.8	1.8	2.2	3.5	5.0	7.0	14.7	45.3
H_2	0.2	0.05	0.7	1.3	0.8	2.1	3.8	6.4
H_4	0.1	0.01	0.02	0.03	0.01	0.05	0.8	1.1
CO	0.4	0.3	1.2	1.3	0.6	1.0	2.8	19.9
CO_2	0.8	0.5	0.4	0.8	2.1	1.8	2.5	15.1
N_2	0.3	0.9	–	–	–	1.0	–	–
SO_2	–	–	–	–	1.3	–	2.6	–

Dose range: 4–12 MGy

also affected by the polymer morphology as seen in the results on amorphous and crystalline PEEK. The G-values of total gas evolution for the aromatic polymers were found to be 10^{-2} to 10^{-3} of those for conventional aliphatic polymers. The low yields of gas evolution from the aromatic polymers are closely correlated to the incorporation of aromatic moieties into polymer backbone which greatly enhances the tolerance to high energy radiation.

These radiation-resistant aromatic polymers containing heteroatoms recently draw much attention from the viewpoint of materials for fusion reactor applications. However, no investigation has been done on radiation effects of these polymers at cryogenic temperatures. Further experiments are required to make the selection of available polymer materials for fusion reactors.

3.4 Polymer Composites

Reinforced polymers with inorganic fillers are known to show greater resistance to radiation than original polymers. Most of organic materials in fusion reactors must be designed to employ organic-matrix composites as it appears very doubtful that conventional polymers can withstand the severe radiation environment for a reasonably long operation time.

Epoxy resins are widely used as organic matrices for composites in insulating and structural applications where maximum strength is required.

Epon 828 resins, which are based on diglycidyl ether of bisphenol A filled with 40 wt.% SiO_2, were irradiated at 4.9 K and tested at 77 K after being warmed up to room temperature. The flexural and compressive strengths of the filled epoxy resins were found to be little affected by a γ-ray dose of 2×10^7 Gy, but to deteriorate significantly after exposure to 1×10^8 Gy [49].

Coltman and Klabunde carried out the cryogenic irradiation of G-10CR and G-11CR which are glass-cloth-reinforced epoxy resins mentioned in the preceding chapter [61]. The flexural tests of specimens were made at 77 K after irradiation at 4.9 K and warm-up to room temperature. The changes in flexural strength are shown in Fig. 9, where G-10CR(BF) means a resin used boron-free

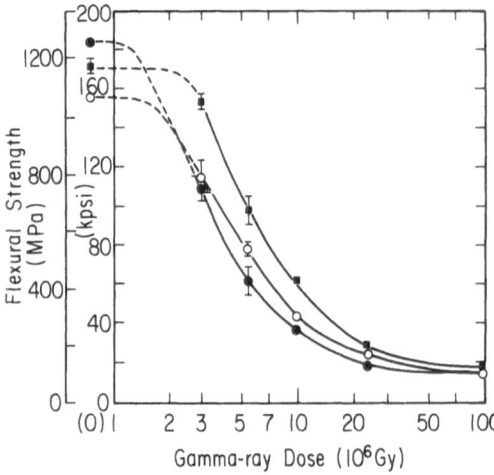

Fig. 9. Radiation effects on the flexural strength of G-10CR, G-10CR(BF) and G-11CR irradiated at 4.9 K. (●) G-10CR; (○) G-10CR(BF); (■) G-11CR

glass-cloth. The strength of G-11CR was reduced by 50% after γ-ray irradiation to about 6.2×10^6 Gy. The strengths of G-10CR and G-10CR(BF) fell to half of their original strengths at 3.5 and 5.5×10^6 Gy, respectively. The difference of damage between G-10CR and G-10CR(BF) is attributed to the additional absorbed dose in the former from the fission of ^{10}B atoms in the glass by the capture of thermal neutrons. The data obtained for compressive strength and linear flexural modulus also show that G-11CR is more radiation resistant than G-10CR. These results are easily understandable in terms of intrinsic radiation tolerance of the matrix in G-11CR. Both visual and low-power stereomicroscopy studies on the flexural specimens of G-10CR and G-11CR suggest that debonding induced by irradiation occurs between the glass fiber and the matrix, which is closely related to the loss in mechanical strength [62]. The experiments on the fatigue resistance of G-11CR revealed that the initial failure of the composite at 77 K occurred as delamination in the notch region and the radiation effect was more pronounced at 77 K than at 295 K [63].

Coltman and Klabunde also examined the radiation effects of glass-fabric filled polyimides called Spaulrad and Norplex NP 530 [50]. Spaulrad is a high-pressure laminate composed of aromatic polyimide resin reinforced with continuous filament E glass-woven fabric 70–71% by weight. Norplex NP 530 is a balanced resin of bismaleimide and aromatic diamines reinforced with E glass-woven fabric 40–60% by weight. Figure 10 shows the changes in flexural strengths of Spaulrad and Norplex at 77 K as a function of γ-ray dose, together with the results of irradiated Vespel. At a dose of 2.4×10^7 Gy, Spaulrad is about five time stronger than G-11CR with tests at 77 K. The flexural strengths of Spaulrad and Norplex at 1×10^8 Gy are reduced to 62% and 70% of their initial values at 77 K, respectively. However, both composites still retain useful strength, indicating that polyimide composites are promising as an insulator of fusion magnets. The stronger nature of Spaulrad than Norplex throughout the

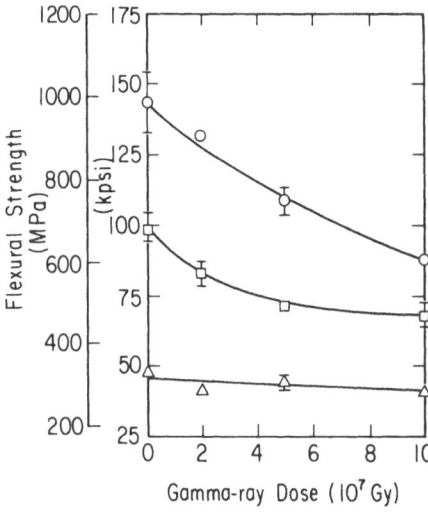

Fig. 10. Radiation effects on the flexural strength on pure and glass-fabric filled polyimides irradiated at 4.9 K. (○) Spaulrad; (□) Norplex; (△) Vespel

whole range of doses is possibly due to the higher content of glass-fabrics in Spaulrad. From the results of flexural and compressive tests, the radiation-induced loss in mechanical strength of the specimens is believed to be caused by the deterioration of the glass-to-polyimide adhesion, because the strength of pure polyimide is little changed for doses examined.

Figure 11 shows the ultimate flexural strength measured at 4.2 K as a function of the absorbed dose in matrix for four kinds of E-glass reinforced epoxy composites, together with a polyimide (polyaminobismaleimide, Kerimid 601) composite, where γ-irradiation of the specimens were carried out at room temperature [64]. The radiation resistance of these composites was found to increase in the order of G-10CR < G-11CR ~ DGEBA/DDM < TGDDM/DDS < polyimide and also observed that TGDDM/DDS composite is superior to the polyimide composite in terms of the mechanical properties at 4.2 K in a dose range up to at least 70 MGy.

The radiation effects on compression fatigue of five kinds of polymer composites, in which two samples were prepared with E-glass cloth (containing B_2O_3) and three samples with S-glass cloth (boron free), were examined by Schmunk et al. [65]. The specimens were irradiated at 325 K in the Advanced Test Reactor of Idaho Falls up to a γ-ray dose of about 3×10^9 Gy and a total neutron flux of about 4×10^{24} n/m^2 and were followed by testing of low-cycle compression fatigue at room temperature. No failure was observed on the three composites containing S-glass (matrix resins: two kinds of epoxy and Kerimid 601) performed with an applied maximum compression stress of 345 MPa for over 1.7×10^5 cycles. In comparison, E-glass reinforced composites (G-10 and G-11CR) showed rapid failure after only a few hundred cycles. The reason is that the total radiation dose to the specimens containing E-glass is considerably higher due to the thermal neutron fission of ^{10}B atoms as mentioned above. The

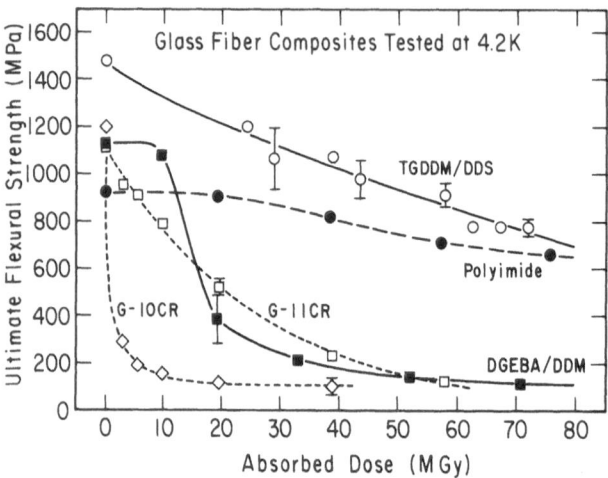

Fig. 11. The ultimate flexural strength at 4.2 K as a function of the absorbed dose in matrix for various polymer composites. (○) TGDDM/DDS (tetraglycidyldiaminodiphenylmethane cured with diaminodiphenylsulfone); (■) DGEBA/DDM (diglycidyl ether of bisphenol A cured with diaminodiphenylmethane); (◇) G-10CR; (□) G-11CR; (●) polyimide composite

same conclusion was recently obtained in the interlaminar shear tests of reactor irradiated polymer composites at 20 K by Nishijima et al. [66, 67]. These results recommend the use of boron-free glass as a reinforced fiber for polymer composites in the intense thermal neutron field.

Neutronic calculations show that fast neutron fluxes at fusion reactor magnet locations are rather high and the energy deposited by neutrons compares to or exceeds that by γ-rays [2]. It is therefore important to establish the characteristics of radiation damage due to fast neutrons in comparison with those due to γ-rays.

Hurley et al. irradiated the epoxy-matrix laminate G-10CR in liquid helium with fast neutrons from the Intense Pulsed Neutron Source (IPNS) at the Argonne National Laboratory [68]. As a result of dosimetry, it is found that the absorbed energy for the epoxy matrix is much greater than for the E-glass fibers, although the neutron flux (1.52×10^{21} n/m², E > 0.1 MeV) is the same for both components of the composites. Neutron irradiation of G-10CR with a dose of 2.5×10^6 Gy resulted in small but significant changes in mechanical properties. Microscopic observation on fracture surfaces of neutron irradiated G-10CR, after 75 K tests, revealed a marked change in appearance with preferential failure near the fibers occurring through the resin. This may be explained in terms of dose effects due to the heterogeneous nature of the composites, where hydrogen-rich organosilane used as a coupling agent makes the interfacial region preferred sites for energy deposition under neutron irradiation. The fast neutron irradiation on G-10CR and G-11CR were also performed by the same research group [69, 70]. As a result of being irradiated at 4.2 K up to a fast neutron flux of 4.1×10^{21} n/m² (E > 0.1 MeV), G-10CR showed a significant reduction in flexural strength (67%) and G-11CR exhibited only a small decrease in strength (12%).

The effects of neutron irradiation on four kinds of cloth-filled polymer composites (filler: E-glass or carbon cloth; matrix: epoxy or polyimide resin) at 5 K and room temperature were also examined with the use of IPNS by Egusa et al. [71, 72]. For the exact evaluation of the neutron dose in polymer composites, the conversion factors from total neutron flux (n/m^2) to absorbed dose (Gy) were estimated by taking into account the range of recoil particles and the linear dimensions of matrix resin and reinforcing fiber. The amount of energy deposition due to recoil particles in a polymer composite was shown to be strikingly affected by the presence of reinforcing fibers, depending on the IPNS neutron spectra and on the kinds of composite materials [71]. All mechanical tests were performed at 77 K after warm-up to room temperature for the specimens irradiated at 5 K. The Young's modulus of these composites scarcely changed during the neutron irradiation up to 3.0×10^{22} n/m^2 (2.1 $\times 10^{22}$ n/m^2, E > 0.1 MeV) at 5 K and 1.6×10^{23} n/m^2 (8.0×10^{22} n/m^2, E > 0.1 MeV) at room temperature. However, the ultimate flexural strength of the composites except the carbon/epoxy composite obviously decreased with increasing neutron flux at both temperatures. No significant difference on the degradation efficiency of the strength was observed between the specimens irradiated at 5 K and those at room temperature. The comparison of results on neutron irradiation with those obtained by γ-irradiation [73] revealed that the radiation sensitivity of the glass/epoxy and glass/polyimide composites is 1.8–2.6 times higher towards neutrons than γ-rays [72]. This indicates that the radiation effects of neutrons are different from those of γ-rays not only as to the energy-deposition distribution in composite specimens but also as to the decomposition efficiency of matrix resins. On the basis of these results, Egusa concluded that the data of simulated-irradiation with γ-rays or accelerated electrons are unreliable as sources of design data for fusion magnets [74, 75].

In order to establish the evaluation on radiation damage of polymer composites, Okada and his coworkers tried to analyze the fracture mode of irradiated composites in both flexural and tensile tests [76]. The change in fracture mode from bending or tensile to shear resulted in the decrement in flexural or tensile strength of the irradiated composites. Thus the essential property governing the deterioration of polymer composites appeared to be the interlaminar shear strength. The difference between neutron and γ-irradiations at both cryogenic and room temperatures were obviously demonstrated by the changes in interlaminar shear strength of glass-cloth-reinforced epoxy composites [77].

Recently Nishijima et al. investigated the radiation effects of three-dimensional glass-fabric reinforced plastics (3DFRP) mentioned in the preceding section, since the interlaminar shear strength of composites was expected to be greatly enhanced by the presence of Z-axis reinforcement [78]. Two kinds of 3DFRP were newly developed and named as ZI-003 and ZI-005 of which the matrices were epoxy and BT resins, respectively [28]. The compressive tests

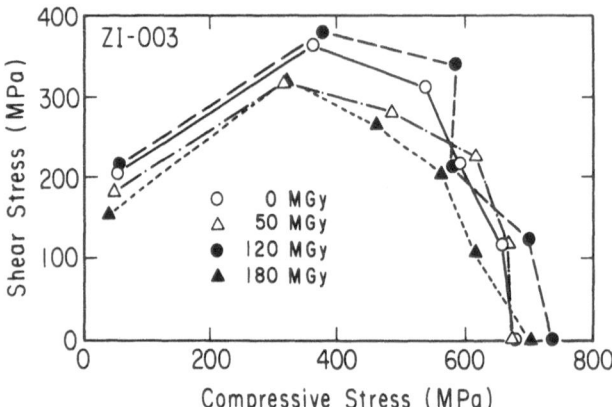

Fig. 12. Effect of electron irradiation on compressive strength under shear stress of 3DFRP measured at room temperature

under shear stress of 20 MeV electron-irradiated specimens were performed at room temperature by using a V-shaped compressive jig. The results obtained for ZI-003 are represented in Fig. 12. It is evident that ZI-003 can withstand complicated stress conditions under severe radiation environments. The specimens of ZI-005 showed themselves to be more radiation-resistant than those of ZI-003.

The mechanical tests for two kinds of 3DFRP (matrix: epoxy and BT resins) together with Spaulrad-S (matrix: bismaleimide) were also carried out at room temperature [79]. The specimens were irradiated to 5×10^7 and 3×10^8 Gy with 35–40% of the total dose of neutrons at the Advanced Technology Reactor. The interlaminar shear strength was found to remain at a high level for all three FRP even after exposure to 3×10^8 Gy. The 3DFRP exhibited almost identical properties to Spaulrad-S at the test with combined shear and compression, but showed higher strength and better radiation resistance than Spaulrad-S.

These results indicate that the newly developed 3DFRP are potentially component materials of fusion reactors. In fact, the research groups of Japan [78] and the US [79] are planning to examine the cryogenic properties of irradiated 3DFRP for insulating materials of fusion magnets.

3.5 Radioactivity of Irradiated Composites

One of the severe problems associated with material selection for use in the presence of neutrons is the production of radioactive nuclides formed by neutron activation. It is therefore important to investigate radioactive species produced from inorganic fillers in composites.

Coltman et al. first measured the radioactivity of several epoxy composites for some weeks following their reactor irradiation [49]. Although the measurements were rather qualitative, they observed that the radioactivity of both

G-10CR and G-11CR was about 20–25 times greater than that of the particle-filled composites such as Stycast 2850 FT and EPON 828.

For the purpose of identifying radioactive nuclides in irradiated fillers, the neutron activation analyses for several samples were carried out by Yamaoka and his coworkers [80]. Figure 13 shows an example of γ-ray spectra on

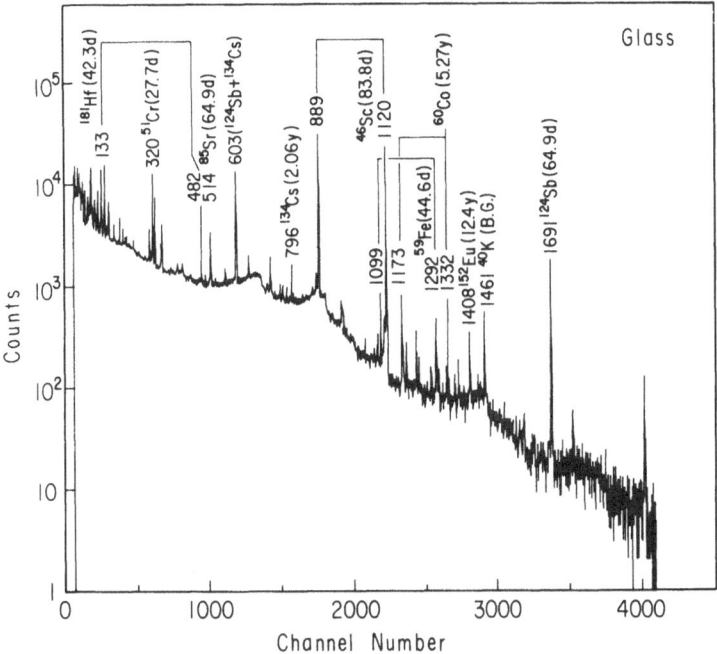

Fig. 13. Gamma-ray spectra on neutron-irradiated E-glass fiber measured after 30 days of cooling

Table 4. Long-lived radioactive nuclides produced by neutron irradiation in inorganic fillers for polymer composites

Nuclide (Half-life)	Filler				
	Glass	Mica	SiO_2	Al_2O_3	SiC
^{46}Sc (83.8 d)	4.23	6.50	0.04	–	–
^{51}Cr (27.7 d)	65.0	1.95×10^3	3.87	0.54	0.66
^{59}Fe (44.6 d)	1.41×10^3	3.97×10^4	9.93	–	0.16
^{60}Co (5.27 y)	2.43	53.7	0.29	0.01	0.03
^{65}Zn (244 d)	–	–	16.5	2.07	3.55
^{85}Sr (64.9 d)	2.22×10^3	–	–	–	–
^{110m}Ag (252 d)	–	–	–	0.02	0.39
^{124}Sb (60.3 d)	18.3	–	21.1	0.37	–
^{134}Cs (2.06 y)	1.83	3.13	–	–	–
^{152}Eu (12.4 y)	0.05	0.82	–	–	–

Concentration: μg/g filler

neutron-irradiated E-glass fiber which was measured after 30 days of cooling in order to eliminate the components of short-lived nuclides. As can be seen in Fig. 13, various kinds of long-lived radioactive species were formed in the irradiated sample of E-glass fiber. The results obtained are summarized in Table 4. In the irradiated mica flake as well as E-glass fiber, a large number of radioactive species were detected. On the other hand, no appreciable amounts of long-lived components were found in the irradiated specimens of SiO_2 powder, Al_2O_3 and SiC fibers. From the viewpoint of neutron-induced radiaoactivity, Al_2O_3 and SiC fibers seem to be desirable as candidate fillers of several composites used in intense neutron fields, although these fibers are much more expensive than commercially available glass fibers.

4 Electrical Properties at Cryogenic Temperatures

The electrical properties of insulators for superconducting magnets are of crucial importance in relation to the operational reliability of fusion reactors [81]. In the present section, the characteristics in original electrical properties of polymers at cryogenic temperatures are briefly introduced and then the effects of radiation on these properties are surveyed.

4.1 Unirradiated Conditions

The electrical breakdown of insulating materials is a very complex phenomenon which is dependent on many factors such as intrinsic, defect-dependent, time-dependent due to corona attack and thermal runaway resulting from dielectric losses. The breakdown processes on polymers were thoroughly reviewed by Ieda [82]. Figure 14 shows examples of the temperature dependence on field strength of various polymers which are carefully measured with DC voltage to eliminate the edge effect. In general, the maximum values of field strength of polymers are observed in the lower temperature region. The values for polar polymers are over 10 MV/cm and are higher than those for nonpolar polymers.

In addition to high breakdown strength, the electrical insulators for super-conducting magnets must have excellent dielectric properties at cryogenic temperatures. Chant reported the results of measurements on dielectric constant and loss tangent (tan δ) for several polymers over the temperature range from 4.2 to 300 K [83]. The variation of dielectric constant of samples as a function of temperature is shown in Fig. 15. The dielectric constants of nonpolar polymers, such as polyethylene, polypropylene and polytetrafluoroethylene, are substantially independent of temperature, whereas those of polar polymers except polyimide decrease by a maximum of 20% as the temperature is reduced. The values of tan δ at the frequency of 75 cps for nonpolar polymers decreased by

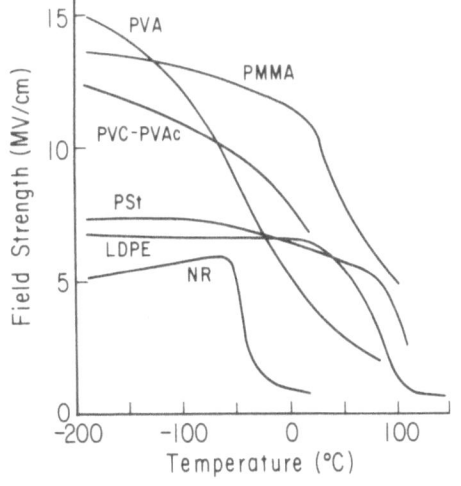

Fig. 14. Temperature dependence on the field strength of various polymers

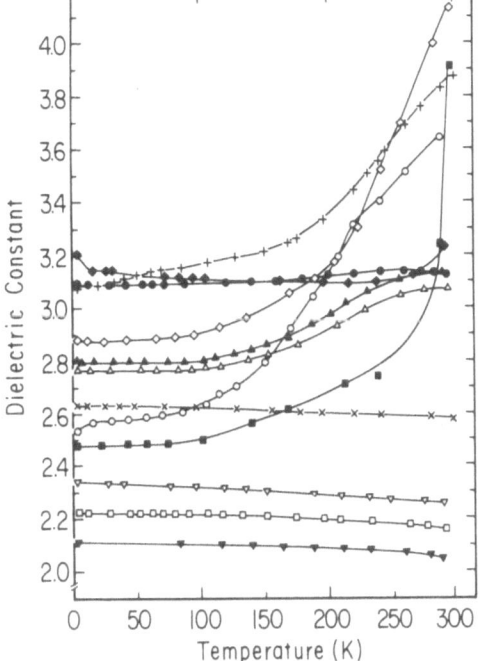

Fig. 15. Temperature dependence on the dielectric constant of various polymers at 75 cps. ($\diamond$) Polyhexamethylene-adipamide 0.005 in; ($\blacklozenge$) Silicone-bonded samica 0.004 in; ($\bigcirc$) Cellulose 0.005 in; ($\bullet$) Polyimide 0.0048 in; ($\square$) Polyethylene 0.005 in; ($\blacksquare$) Rubber hydrochloride 0.002 in; ($\triangle$) Polyethylene terephthalate (Melinex) 0.005 in; ($\blacktriangle$) Polyethylene terephthalate (Mylar) 0.002 in; ($\triangledown$) Polypropylene 0.010 in; ($\blacktriangle$) Polytetrafluoroethylene 0.0075 in; ($\times$) Polystyrene 0.005 in; ($+$) Aromatic polyamide 0.0027 in (1 in = 2.54 cm)

one or two orders of magnitude between room temperature and 4.2 K. The lowest tan δ values for polyethylene and polypropylene were about $2\text{--}3 \times 10^{-5}$ at 4.2 K, compared with around 10^{-4} for polar polymers. The tan δ values for polar polymers also decreased considerably at cryogenic temperature, but they were found to remain frequency dependent even at 4.2 K. The degree of crystallinity of polymers is known to affect significantly the magnitudes of the loss tangent. An example of the effect of crystallinity on tan δ in the case of PEN is shown in Fig. 16 [84]. The peak height of tan δ around 230 K decreases with increasing the degree of crystallinity, indicating that this peak is closely associated with the segment motion in an amorphous part.

The effect of cryogenic temperatures on the volume resistivity and electrical strength of G-10CR and G-11CR epoxy laminates was first reported by Kasen et al. [21], as summarized in Table 5. The electrical strength was found to be independent of both temperature and composite type. The effect of temperature on the volume resistivity was similar for the two composites, increasing 1.5 to 2 orders of magnitude on cooling from 295 to 4 K.

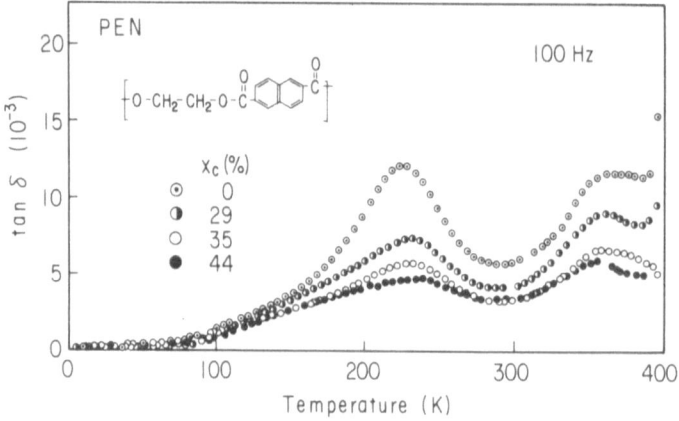

Fig. 16. Effect of crystallinity on tan δ of PEN as a function of temperature

Table 5. Electrical properties of G-10CR and G-11CR

Temperature (K)	Thickness (cm)	Volume resistivity ($\Omega \cdot$ cm)	Voltage stress at breakdown (kV/mm)
		G-10CR	
295	0.030		48.4
	0.036	8.9×10^{15}	45.7
	0.051	9.3×10^{14}	
77	0.051	1.5×10^{17}	
4	0.036	4.0×10^{17}	48.4
	0.051	4.1×10^{17}	
		G-11CR	
295	0.030	7.3×10^{15}	
	0.051	1.3×10^{15}	
77	0.051	6.1×10^{16}	
4	0.030	2.0×10^{17}	48.0
	0.051	2.0×10^{17}	

It is generally considered that the electrical properties of polymers and composites tend to increase their values to advantageous regions on cooling to cryogenic temperatures.

4.2 Irradiated Conditions

Both transient and permanent changes in the electrical properties of polymers are induced by the action of radiation [85]. The transient effects are sensitively affected by the kind of radiation and its intensity, because the magnitude and the distribution of any space charge in materials depend on generation and recombination rates of charged species. Although the transient effects are currently interesting topics, only the permanent effects by radiation on electrical properties are taken up in this section from the viewpoint of practical uses.

Banford et al. studied the radiation effects on electrical properties of low-density polyethylene (LDPE) at 5 K with the use of a ^{60}Co gamma source and a thermal nuclear reactor [86]. They reported that both the electrical conductivity and the dielectric breakdown strength of LDPE at 5 K were not significantly affected by radiation absorbed doses up to 10^5 Gy, but an erratic pulse activity under high applied fields was observed in the sample irradiated at 10^6 Gy.

In the case of Mylar, the electrical breakdown and the dielectric constant showed little change after the irradiation of 7×10^6 Gy at 5 K, although all mechanical properties fell down to unusable levels [87].

Figure 17 represents the temperature dependence of the breakdown strength in unirradiated and irradiated PPS films [52]. It appeared that the reactor irradiation of 8×10^6 Gy at 20 K gave no significant effect on the breakdown strength of PPS films over a wide temperature range.

The radiation effects on dielectric properties of an epoxy resin (Epilox EG 34 with aromatic amine hardener Nr 105) were studied by Jahn et al. with electron,

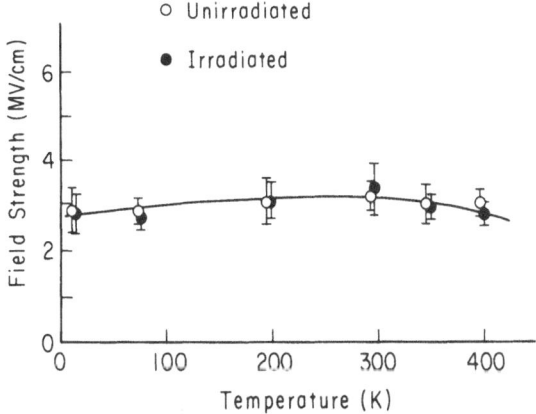

Fig. 17. Temperature dependence on the breakdown strength of PPS. (○) control; (●) irradiated up to 8×10^6 Gy at 20 K

fast and thermal neutron irradiations at room temperature [88, 89]. All of non-irradiated and irradiated samples showed a maximum peak of tan δ at about 245 K with the frequency of 10 KHz, being connected to a vibration of hydroxy groups involved. In the electron irradiation, the dielectric constant and the peak height of tan δ decreased with increasing an absorbed dose. After the irradiation of 1×10^7 Gy, the decreased values of dielectric constant and tan δ at 245 K were 9% and 26%, respectively. Similar effects were also observed in the thermal neutron irradiation in which both values with a thermal neutron flux of 2.6 $\times 10^{21}$ n/m^2 were 27% and 16%, respectively. In the case of the sample after fast neutron irradiation of 1.0×10^{20} n/m^2, no apparent change on dielectric properties was recognized between the original sample and the irradiated one. The samples irradiated by 10^{20} n/m^2 of fast neutrons, however, were destroyed by the recoil protons. The effect of γ-irradiation on electrical properties of epoxy resins were also investigated systematically by Wu et al. [90].

There is little data on the change in electrical properties of composites as a function of absorbed dose. Table 6 is the summarized results for epoxy composites, together with those for Nomex and Kapton, obtained by Coltman et al. [49], where electrical measurements were carried out at room temperature after irradiation at 4.9 K. Both resistivities and dielectric breakdown strengths of all composite specimens were confirmed to remain constant up to a γ-ray dose of 1×10^7 Gy, at which the mechanical strengths of these composites were remarkably reduced by over 50%. The resistivity of Stycast FT was virtually unaffected by the irradiation up to a dose of 1.0×10^8 Gy, whereas that of Epon 828 was observed to decrease with the irradiation over 0.24×10^8 Gy. Both G-10CR and G-11CR exhibited little change in resistivity with the irradiation of 0.24 $\times 10^8$ Gy, but significant decreases with the irradiation of 1.0×10^8 Gy. The drastic decrease in dielectric breakdown for epoxy composites except Epon 828 occurred between 0.24 and 1.0×10^8 Gy, but these values were still held in a

Table 6. Electrical measurements of epoxy composites and polymer films at room temperature after irradiation at 5 K*

Dose (10^8 Gy)	Stycast 2850 FT	EPON 828	G-10CR	G-11CR	Nomex 410	Kapton H
			Resistivity (10^{13} Ω·m)			
Control	0.24	20	8.2	4.4	3.3	20
0.24	0.25	5.5	8.5	2.8	2.2	29
1.0	0.27	8.1	0.64	0.14	2.0	22
			Dielectric breakdown strength** (kV/mm)			
Control	28[2]	33[0]	23[0]	24[0]	36[2]	66[0]
0.24	31[0]	34[0]	23[0]	23[0]	38[2]	66[0]
1.0	8[3]	31[1]	8[3]	10[3]	36[2]	68[0]

* Each result was the average of three measurements.
** The superscripts indicate the number of samples exhibiting volume breakdown

Table 7. Electrical resistivity measurements on polymer composites

Material	Resistivity	
	Unirradiated ($\Omega \cdot$ cm)	Irradiated ($\Omega \cdot$ cm)
G-10	1.1×10^{16}	3.8×10^{7}
G-11CR	4.1×10^{15}	5.6×10^{9}
KERIMID-601	1.4×10^{15}	7.9×10^{11}
DGEBA	8.9×10^{15}	1.6×10^{12}
TGPAP	2.2×10^{15}	6.6×10^{11}

usable range. On the other hand, both Nomex 410 and Kapton H showed little change in their electronic properties even with irradiation as high as 1.0 $\times 10^8$ Gy.

Tucker et al. reported the results of electrical property measurements on fast neutron irradiated GFRP [69]. G-10CR, G-11CR and two polyimide based GFRP (Spaulrad and NP-530) were irradiated at 4.2 K up to a total dose of 4.1 $\times 10^{21}$ n/m^2 (E > 0.1 MeV) for DC resistivity measurements and that of 2.6 $\times 10^{21}$ n/m^2 for dielectric breakdown strength tests. Although G-10CR exhibited a large reduction in mechanical strength with the fast neutron irradiation, both DC conductivity and dielectric strength on each sample revealed little or no pattern of degradation at all levels of radiation exposure.

Table 7 shows the results of electrical resistivity measurements for five insulator materials irradiated heavily at 325 K in the Advanced Test Reactor [65]. Irradiation conditions were the same as those for mechanical tests mentioned in the preceding section. The remarkable decreases of resistivity were observed in both G-10 and G-11 CR which were made with E-glass cloth. The other three specimens containing S-glass cloth showed fairly good tolerance of electrical resistivity to radiation.

In general it appears that the electrical properties of organic materials practically do not change, as long as the mechanical properties of materials withstand the action of radiation.

5 Concluding Remarks

The research and development efforts on polymer materials for fusion reactors have been intensified in recent years. Some polymers and composites are able to withstand the radiation doses in excess of 10^8 Gy even at cryogenic temperatures. Furthermore, international research projects on organic insulators used for fusion magnets are currently in progress. As one of the important subjects, the combined effects of intense radiation and thermal cycling are being tested.

Although it is unlikely that a large-scale fusion reactor will be operational within this century, the data described here are certainly significant for improving conceptual designs of prototypic fusion reactors. It should be emphasized, however, that these data may only be used as a guide for the selection of fusion reactor components, since these were obtained using so-called laboratory systems. Much uncertain factors on actual environmental conditions of fusion reactors would affect various properties of polymer materials. In order to judge the performance of polymer materials for fusion reactors, the required measurements must be extended to include the engineering data on large-scale specimens. Further investigation is needed to develop the new type of polymer materials which can endure the severe conditions in fusion reactor environments.

6 References

1. Akiyama M (ed) (1991) Design technology of fusion reactors. World Scientific, Singapore (Series in Theoretical and Applied Mechanics, vol 6)
2. Abdou MA (1978) J Nucl Mater 72: 147
3. Brown BS (1981) J Nucl Mater 97: 1
4. van der Klein CAM (1975) RCN-240, Reactor Centrum Nederland
5. Hay RD, Rapperport EJ (1976) TID-27124, Oak Ridge National Laboratory, Tennessee
6. Phillips DC (1978) AERE-R8923 Harwell
7. Clinard FW, Hurley GF (1981) J Nucl Mater 103/104: 705
8. Coltman Jr RR (1982) J Nucl Mater 108/109: 559
9. Hurley GF, Coltman Jr RR (1984) J Nucl Mater 122/123: 1327
10. Woodward AE (1966) J Polym Sci C 14: 89
11. Nielsen LE (1975) Mechanical properties of polymers and composites. Marcel Dekker, New York
12. Armeniades CD, Baer E (1971) J Polym Sci Pt A-2 9: 1345
13. Armeniades CD, Kuriyama I, Roe JM, Baer E (1967) J Macromol Sci Phys B1: 777
14. Reed RP, Schramm RE, Clark AF (1973) Cryogenics 13: 67
15. Bernier GA, Kline DE (1968) J Appl Polym Sci 12: 593
16. Butta E, Petris SD, Pasquini M (1969) J Appl Polym Sci 13: 1073
17. Ahlborn K (1988) Cryogenics 28: 234
18. Yamaoka H, Miyata K (1986) Adv Cryog Eng (Mater) 32: 161
19. Chamberlain DE (1964) Adv Cryog Eng 9: 131
20. Hertz J (1965) S P E Journal 21: 181
21. Kasen MB, MacDonald GR, Beekman Jr DH, Schramm RE (1980) Adv Cryog Eng (Mater) 26: 235
22. Hamelin J (1980) Adv Cryog Eng (Mater) 26: 295
23. Fukushi K, Nagai M, Kamata Y, Kadotani K (1984) Mechanical properties of low thermal contraction GFRP. In: Hartwig G, Evans D (eds) Nonmetallic materials and composites at low temperature, vol 3. Plenum, New York, p 187
24. Nishijima S, Wang Y-A, Okada T, Uemura T, Hirokawa T, Yasuda J (1988) Adv Cryog Eng (Mater) 34: 59
25. Okuyama H, Nishijima S, Okada T, Iwasaki Y, Yasuda J, Hirokawa T (1990) Adv Cryog Eng (Mater) 36B: 901
26. Iwasaki Y, Nishijima S, Yasuda J, Hirokawa H, Okada T (1990) Adv Cryog Eng (Mater) 36B: 969
27. Yasuda J, Hirokawa T, Iwasaki Y, Nishijima S, Okada T (1990) Adv Cryog Eng (Mater) 36B: 985

28. Iwasaki Y, Yasuda J, Hirokawa T, Noma K, Nishijima S, Okada T (1990) Cryogenics 31: 261
29. Kasen MB (1975) Cryogenics 15: 327
30. Kasen MB (1975) Cryogenics 15: 701
31. Kasen MB (1981) Cryogenics 21: 235
32. Kasen MB (1982) Adv Cryog Eng (Mater) 28: 165
33. Hartwig G (1982) Adv Cryog Eng (Mater) 28: 179
34. Hartwig G (1984) Cryogenics 24: 639
35. Hartwig G (1986) Cryogenic properties. In: Encyclopedia of polymer science and engineering, 2nd edn, vol 4, John Wiley, New York, p 450
36. Reed RP, Fickett FR (eds) (1990) Adv Cryog Eng (Mater), vol 36, Pt B, pp 787–1036
37. Hartwig G, Evans D (eds) (1991) Cryogenics, vol 31, No. 4
38. Beynell P, Maier P, Schönbacher H (1982) Compilation of radiation damage test data - - - Part III: Materials used around high-energy accelerators, CERN 82-10, European Organization for Nuclear Research, Geneva
39. Schönbacher H (1985) Modern Plastics 62: 64
40. Clough R (1988) Radiation-resistant polymers. In: Encyclopedia of polymer science and engineering, 2nd edn, vol 13. John Wiley, New York, p 667
41. Van de Voorde MH (1971) IEEE Trans Nucl Sci 18: 784
42. Van de Voorde MH (1973) IEEE Trans Nucl Sci 20: 693
43. Evans D, Morgan JT, Stapleton GB (1971) Internal Report RHEL/R 220, Rutherford Laboratory, Chilton
44. Evans D, Morgan JT (1982) Epoxide resins for use at low temperature. In: Hartwig G, Evans D (eds) Nonmetallic materials and composites at low temperatures, vol 2. Plenum, New York, p 73
45. Takamura S, Kato T (1980) Cryogenics 20: 441
46. Nishijima S, Ueta S, Okada T (1981) Cryogenics 21: 312
47. Morgan JT (1970) Internal Report RHEL/R 196, Rutherford Laboratory, Chilton
48. Evans D, Morgan JT (1982) Adv Cryog Eng (Mater) 28: 147
49. Coltman Jr RR, Klabunde CE, Kernohan RH, Long CJ (1979) ORNL/TM-7077, Oak Ridge National Laboratory, Tennessee
50. Coltman Jr RR, Klabunde CE (1981) J Nucl Mater 103/104: 717
51. Yamaoka H, Miyata K (1989) Preprints of 32nd disc meeting on radiat chem Jpn. 19–20 Oct 1989. Hiroshima, p 57
52. Yamaoka H, Miyata (1983) J Nucl Mater 133/134: 788
53. Bell VL, Pezdirtz GF (1983) J Polym Sci Polym Chem Ed. 21: 3083
54. Havens SJ, Bell VL (1986) J Polym Sci Pt A, Polym Chem 24: 901
55. Sasuga T, Hayakawa N, Yoshida K, Hagiwara M (1985) Polymer 26: 1039
56. Sasuga T, Hagiwara M (1987) Polymer 28: 1915
57. Sasuga T (1988) Polymer 29: 1562
58. Hedrick JL, Mohanty DK, Johnson BC, Viswanathan R, Hinkley JA, McGrath JE (1986) J Polym Sci Polym Chem Ed 23: 287
59. Lewis DA, O'Donnell JH, Hedrick JL, Ward TC, McGrath JE (1988) ACS Symp Ser 381: 252
60. Hegazy EA, Sasuga T, Nishii M, Seguchi T (1990) Preprints of 33rd disc meeting on radiat chem Jpn. 8–9 Oct 1990. Sendai, p 131
61. Coltman Jr RR, Klabunde CE (1983) J Nucl Mater 113: 268
62. Klabunde CE, Coltman Jr RR (1983) J Nucl Mater 117: 345
63. Korukonda B, Conway Jr JC, Queeney RA, Diethorn WS (1983) J Nucl Mater 115: 197
64. Egusa S, Nakajima H, Oshikiri M, Hagiwara M, Shimamoto S (1986) J Nucl Mater 137: 173
65. Schmunk RE, Miller LG, Becker H (1984) J Nucl Mater 122/123: 1381
66. Nishijima S, Okada T, Hirokawa T, Yasuda J, Iwasaki Y (1991) Cryogenics 31: 273
67. Okada T, Nishijima S, Nishiura T, Miyata K, Yamaoka H (1992) Adv Cryog Eng (Mater)
68. Hurley GF, Fowler JD, Rohr DL (1983) Cryogenics 23: 415
69. Tucker DS, Clinard Jr DS, Hurley GF, Fowler JD (1985) J Nucl Mater 133/134: 805
70. Tucker DS, Hurley GF, Kennedy JC (1986) Mechanical properties of three candidate organic insulator materials for fusion reactors. In: Hartwig G, Evans D (eds) Nonmetallic materials and composites at low temperature, vol 3. Plenum, New York, p 21
71. Egusa S, Kirk MA, Birtcher RC (1987) J Nucl Mater 148: 43
72. Egusa S, Kirk MA, Birtcher RC (1987) J Nucl Mater 148: 53
73. Egusa S, Hagiwara M (1986) Cryogenics 26: 417
74. Egusa S (1988) J Mech Behv Mater 1: 1
75. Egusa S (1991) Radiat Phys Chem 37: 147

76. Okada T, Nishijima S, Yamaoka H (1986) Adv Cryog Eng (Mater) 32: 145
77. Nishijima S, Okada T, Miyata K, Yamaoka H (1988) Adv Cryog Eng (Mater) 34: 35
78. Nishijima S, Nishiura T, Okada T, Hirokawa T, Yasuda J, Iwasaki Y (1990) Adv Cryog Eng (Mater) 36B: 877
79. McManamy TJ, Kanemoto G, Snook P (1991) Cryogenics 31: 277
80. Yamaoka H, Matsushita R, Miyata K, Nakayama Y (1984) Preprints 1st Intern Conf Fusion React Mater. 3–6 Dec 1984. Tokyo, p 268
81. Banford HM (1984) Electrical insulation and fusion reactors. In: Lewins J, Becker M (eds) Advances in nuclear science and technology, vol 16. Plenum, New York, p 1
82. Ieda M (1980) IEEE Trans Electr Insul EI-15: 205
83. Chant MJ (1967) Cryogenics 7: 351
84. Yano O, Mandai Y, Nakamura M, Soen T, Yamaoka, Miyata K (1989) Polym Preprints Jpn 38: 3425
85. Hedvig P (1972) Electrical conductivity of irradiated polymers. In: Dole M (ed) The radiation chemistry of macromolecules, vol 1. Academic, New York, p 127
86. Banford HM, Frame RI, Tedford DJ (1986) J Nucl Mater 141/143: 410
87. Takamura S, Kato T (1981) J Nucl Mater 103/104: 729
88. Jahn K, Jäckel M, Meyer W (1983) Cryogenics 23: 160
89. Jahn K, Jäckel M, Brunner K (1983) Cryogenics 23: 667
90. Wu S, Gedeon S, Fouracre RA, Tedford DJ (1988) J Nucl Mater 151: 140

Editor: S. Okamura
Received December 11, 1991

Author Index Volume 101–105

Subject Index

Springer-Verlag und Umwelt

Als internationaler wissenschaftlicher Verlag sind wir uns unserer besonderen Verpflichtung der Umwelt gegenüber bewußt und beziehen umweltorientierte Grundsätze in Unternehmensentscheidungen mit ein.

Von unseren Geschäftspartnern (Druckereien, Papierfabriken, Verpakkungsherstellern usw.) verlangen wir, daß sie sowohl beim Herstellungsprozeß selbst als auch beim Einsatz der zur Verwendung kommenden Materialien ökologische Gesichtspunkte berücksichtigen.

Das für dieses Buch verwendete Papier ist aus chlorfrei bzw. chlorarm hergestelltem Zellstoff gefertigt und im ph-Wert neutral.